OCEAN WORLDS

The story of seas on Earth and other planets

JAN ZALASIEWICZ
AND MARK WILLIAMS

OXFORD
UNIVERSITY PRESS

Great Clarendon Street, Oxford, OX2 6DP,
United Kingdom

Oxford University Press is a department of the University of Oxford.
It furthers the University's objective of excellence in research, scholarship,
and education by publishing worldwide. Oxford is a registered trade mark of
Oxford University Press in the UK and in certain other countries

First published 2014
First published in paperback 2017

Impression: 1

Published in the United States of America by Oxford University Press
198 Madison Avenue, New York, NY 10016, United States of America

British Library Cataloguing in Publication Data
Data available

Library of Congress Cataloging in Publication Data
Data available

ISBN 978–0–19–967288–2 (Hbk.)
ISBN 978–0–19–967289–9 (Pbk.)

Printed in Great Britain by
Clays Ltd, St Ives plc

OCEAN WORLDS

an Zalasiewicz teaches and researches geology at the University of eicester, and previously was a field geologist and biostratigrapher at he British Geological Survey. His interests range from the early alaeozoic world of half a billion years ago, to the geology of the resent day and the future. He has served with the Palaeontographical ociety and the Geological Society of London, and is now Chair of the nthropocene Working Group of the International Commission on tratigraphy.

ark Williams is a palaeontologist who teaches the geological istory of climate change at the University of Leicester. He has worked s a field geologist for the British Geological and British Antarctic urveys, and served on the council of the Palaeontographical Society oth as an Editor and Vice-President. Currently he is a member of the nthropocene Working Group of the International Commission on tratigraphy.

Together they have co-authored *The Goldilocks Planet: The four billion year story of Earth's climate* (OUP, 2012).

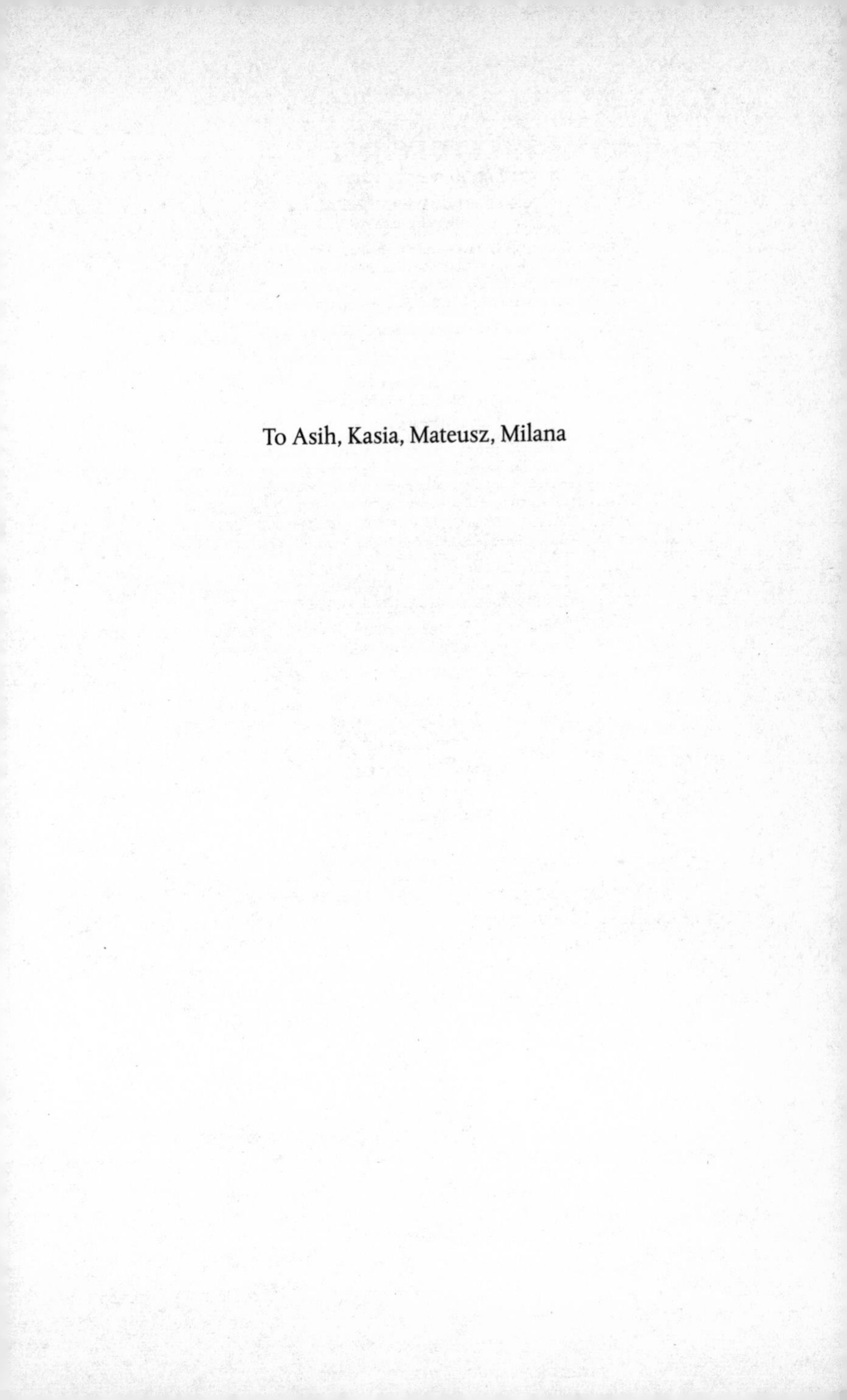

To Asih, Kasia, Mateusz, Milana

Preface

As we walk and drive through our wide landscapes, it can be hard to remember that most of the Earth's surface is taken up by another world. It is a liquid world, kilometres deep, within which we cannot breathe, although many other organisms can. For most of our history it was, except for its surface and shallowest reaches, inaccessible and invisible: the distant stars seemed within closer reach. One could imagine—and people did—monsters as fabulous as the mighty kraken, and ancient submerged civilizations in those deeps.

It is about a century and a half since humans began to seriously explore its depths. In the last few decades it has become clear that our own oceanic world can at times surpass fiction and fable—and only in the last few years has it become clear that, in the distant reaches of outer space, there are many other ocean worlds of astonishing variety. Then there is the impact of the Earth's oceans on our own land-based lives, which is seen to be ever greater the more it is examined—although for some time we have known that the oceans are the source of life-giving rains, and of fish for our dinner plates too.

This book is an attempt to give some picture of the workings of ocean worlds—our own, obviously—and others too, through time and across space. We are geologists and palaeontologists, and are acutely aware that the strata that we measure and the fossils that we glean from them largely represent petrified relics from the Earth's ancient oceans. Being geologists too (and not astronomers, or physicists, or oceanographers) we are acutely aware that we have had to venture far from our familiar territory in order to make this canvas as wide as we wish it to be.

Hence, we are grateful to many people for helping to produce this narrative. Latha Menon, as ever, has been a skilled and supportive editor, while her colleagues at Oxford University Press, notably Emma Ma, have helped keep this whole project on track. John Bridges and Euan Nisbet have provided splendidly thoughtful and useful reviews. We are grateful to images supplied by Ryszard Kryza, Ove Hoegh-Guldberg, Julian Whimpenny, Derek Siveter, Ron Blakey, David Siveter, Jean Vannier, John Bridges, Jon Sykes, Hou Xianguang, Nick Butterfield, Thijs Vandenbroucke, and Latha Menon. We thank Lewis Dartnell, too, for stimulating discussion, Jon Sykes for talking us through the various spacecraft used to image deep space, and our many colleagues down the years, from whom we have learnt pretty much everything that we know: Adrian Rushton and the late Dick Aldridge, in particular, have provided constant inspiration. Our families deserve special thanks, as always, for putting up with the many hours sacrificed to this endeavour.

Contents

List of Illustrations

List of Plates

1

Water in the Cosmos

It was 12 billion years ago, in a far corner of the early universe. A black hole, 20 billion times more massive than our Sun, had formed. It consumed everything around it. Gas, dust, and early-formed star systems fell into its embrace, and the energy of a thousand trillion suns was released in the appallingly brilliant light of a quasar.

What happens there now, in this time, we do not know, and can never know. Perhaps the black hole has eaten all it can, grown yet larger, and now sleeps. It was all a long time ago. We on Earth can, only now, glimpse that example of ancient cosmic gluttony because its light has finally reached us, having taken 12 billion years to travel across the intervening space. It is an image of distant history, fossilized in light beams.

Astronomers call it APM 08279+5255, a title that reveals much about the almost limitless number of objects in the visible universe (we have long ago run out of names such as Arcturus and Proxima Centauri). That hellish light, though, early in its travels (over the first 100 years or so) was lighting up a material that is familiar to us, and that, more than anything else, *is* us.

Water. Permeating that region of the early universe around APM 08279+5255 was a mass of water vapour sufficient to make no less than 140 trillion Earth oceans.[1] It is spread thinly, to be sure, being some 300 trillion times less dense than the Earth's atmosphere. And it is cold, at −53 degrees Celsius, all this being worked out by the extraordinarily precise measuring capabilities of modern spectrometric arrays. But that is still much hotter and denser than the water that is scattered, say, across the Milky Way. It shows that water—the stuff of our oceans—has been virtually everywhere, from close to the beginning of time itself. It is not some kind of cosmic rarity that our Earth, by some quirk, happens to have in abundance.

How is that water made, though, in the depths of the universe?[2] It is, of course, made up of two atoms of hydrogen conjoined with an atom of oxygen: H_2O, the first chemical compound that any child learns at school. That does not mean it is simple. There are deceptive subtleties in that three-atom molecule, as we will see, which make possible all kinds of things, such as oceans, icecaps—and life itself. Nevertheless, one can start with the basic ingredients and go on from there.

With hydrogen it is straightforward. Or straightforward, at least, if we close our eyes to the mysteries that surround the origin of the universe, in the Big Bang, some 13.8 billion years ago. Hydrogen was the main material product of the Big Bang. More precisely, it is the main material product that we can currently observe. There is out there a very great amount of 'dark matter', which must exist, for it holds the galaxies together. Dark matter dwarfs the total amount of normal matter in the universe, but otherwise it remains quite mysterious.[3]

The Big Bang, at its very beginning, was far too hot (at many trillions of degrees Celsius) for hydrogen—or any other kind of atom—to exist. But, when the primordial mass of subatomic particles had stretched and cooled sufficiently, when the universe had grown from nothing to the size of our solar system now, about a millionth of a second into the Big Bang,[4] protons and neutrons began to appear. A single proton is the nucleus of a hydrogen atom—but temperatures (at ten thousand million degrees) were still too high for an electron to fall into orbit around a proton to complete the building of that atom.

That only took place a little later—about 300,000 years later. Then, the universe was about a thousandth of its size today (i.e. already stupendously big) and much cooler, although still hot—about 3,000 degrees Celsius—by human standards. As far as normal matter was concerned, that was largely that. There was also quite a lot of helium—about 25 per cent—for in the furnace of the early Big Bang two protons had occasionally collided violently and precisely enough to combine to form a helium nucleus, the protons managing to overcome their natural mutual repulsion for the short-range 'strong' nuclear forces to bind them together. Occasionally, a third proton was added, to make up a lithium nucleus. These three elements, initially, made up the universe.

So, in those early years of the cosmos there was no possibility of water, because there was no oxygen. Nor was there any sodium, or chlorine, or potassium, or magnesium to make that water salty. There were simply those three primordial elements, swirling in the outrushing clouds of the early universe.

The birth of oxygen and of the other elements of the periodic table required the births—or more precisely the deaths—of the first stars. These first stars began to form as the densest parts of the clouds collapsed under their own weight, creating spinning spheres of gas. These became so tightly compressed at their centres that hydrogen atoms fused to produce yet more helium atoms, releasing the energy that makes stars hot and bright.

Helium is not the only product of the fusion process in the heart of a star. Other elements are made as nuclei continue to combine. This may be either in 'normal' burning processes in large stars, or as the hydrogen supply of any star, large or small, begins to be exhausted. Helium and larger nuclei then fuse to produce nuclei up to the size of iron, releasing yet more energy. In a large star, that process ends in the catastrophic explosion of a supernova, the source of all elements heavier than iron. Depending on the star, and where it is in its history, there are different pathways that lead to different elements.

Oxygen happens to be at the end of some of the most common of these nuclear fusion pathways, and so has, over time, become—by far—the next most common element after hydrogen and helium. Indeed, its total amount in the universe is probably equal to the total amount of all of the other elements (other than hydrogen or helium) put together. The universe, hence, is oxygen-rich. It makes up, for example, almost a third of the whole of the Earth by mass, almost all of it bound up in minerals, particularly in the silicates of the mantle and crust. It is, however, a ferociously reactive element, which is why free oxygen is a rarity in planetary atmospheres—the only[5] exception we know of being the Earth.

When, in the history of the universe, did that oxygen begin to link with what is left of the dying star's hydrogen to form water, in the outrushing material of the nova or supernova? This process is usually assumed to have begun hundreds of millions of years into the history of the universe, with the emergence of the first stars in the billowing gas clouds. But it might, perhaps, have started *much* earlier.

The Harvard astrophysicist Abraham Loeb, in 2013, produced a provocative calculation that suggested that the first stars and planets might have begun forming just 15 million years into the history of the universe.[6] At this time, the temperature of the afterglow of the Big Bang would have cooled to somewhere between 0 and 30 degrees Celsius (it has now cooled to just 3 degrees above absolute zero at −273.15 degrees Celsius). Therefore, Loeb suggested, any planets that might have formed could have had liquid water on their surfaces—and so could potentially have been habitable, even if they were far from their parent star.

This would emphatically have been a weird universe, with matter packed into it a million times more densely than today. Nowhere would have been really *cold*, in that all-round warm blanket. It was a short-lived state, though. After only a few million years the background temperature of the whole universe dropped below zero, so there was not much chance for complex life to emerge before the newly minted waters—the *hypothesized* newly minted waters—nearly everywhere, froze.

In today's familiar, vastly more dilute universe, does an exploding star literally expel steam? In reality, conditions around a nova or (especially) a supernova are mostly too hot for that. The expelled gas is a high-velocity plasma of separate ions. The reactions must mostly take place later, in distant regions, as the gases expand and cool. This is mysterious territory, where the chemical reactions of outer space occur. It needs the most ingenious of modern instrumentation to detect it. Fortunately, such equipment has been devised.

Herschel's Gift

A few years ago, something that looked like a large and eccentrically sculpted trash can was launched into space. It was carefully steered towards one of the points, 1.5 million kilometres away from Earth, where the gravitational forces of the Earth and the Sun are balanced by the centrifugal force of the Earth's passage around the Sun. Such a point in space is sometimes referred to as a gravitational well, and spacecraft can rest there without needing to use too much energy. It is called a Lagrange point, named after their discoverer, Joseph-Louis Lagrange. He was a mathematician so gifted and so temperamentally averse to controversy that he survived all the turmoil of the revolutionary years in France, being awarded honours and distinctions by Louis XVI; then, after Louis succumbed to Madame Guillotine, by the revolutionary government; and then, for good measure, by Napoleon Bonaparte.

This curious object is one of the great triumphs of modern astronomical research. It is the Herschel Space Observatory (Fig. 1), planned and built by the European Space Agency over 30 years, and launched in May 2009. It was destined to be operational for only three years, for the delicate instruments were cooled to just above absolute zero by a small supply of superfluid helium which could not be replaced (so far out in space, no one can make repairs or carry out maintenance). Nevertheless, with a single mirror 3.5 metres across, larger than that of the Hubble Space Telescope, the far-seeing eye of Herschel provided scientists with enough data to occupy them for many years.[7] It finally ran out of its liquid helium supply in 2013.

The singular Herschel Space Observatory was named after two Herschels: brother and sister. There was Sir William Herschel, a German-born musician and soldier who took refuge in England after he was on the losing side of a battle in the Seven Years War. He adapted quickly, in society and professionally. A virtuoso of oboe, violin, and harpsichord, he became director of the Bath Orchestra. His proficiency in music led to an interest in mathematics and, encouraged by his friend Nevil Maskelyne, the Astronomer Royal, in lenses and telescopes. Assiduous in astronomy as in music, he built some of the best reflecting telescopes of his day, including the '40 foot' telescope used to locate Enceladus, the icy moon of Saturn (see Chapter 9). He discovered Uranus and binary star systems, and compiled extensive catalogues of 'nebulae' (galaxies, although they were not recognized as such then) and of star systems.

FIG. 1. An artistic impression of the European Space Agency's Herschel Infrared Telescope, fixed in space some 1.5 million kilometres distant from the Earth.

William's sister Caroline became an astronomer too, through childhood misfortune. Stricken successively with smallpox and typhus as a child, scarred, stunted and deemed unmarriageable, she became a family servant. William rescued her from this drudgery (their mother did not believe in education for women). At his household in England, he taught her music (she became an accomplished singer) and mathematics. She began to help him in his astronomical work, came to be as skilled as her brother, and helped educate another Herschel, William's son John, who grew to become a celebrated astronomer in his own right.

Caroline Herschel discovered several comets, which we now know to be space-borne dirty snowballs. It is apt, therefore, that the Herschel Observatory searched for water in deep space. Earth-bound observatories are hindered by having to peer through an atmosphere that itself is charged with water vapour—but not so the Herschel. Its detectors were set to the infrared, to observe the cool regions of deep space. It found water vapour in the spectra of colliding galaxies, and (perhaps the feedstock for comets) in the disc of a young star, while VY Canis Major, a massive dying star, revealed an envelope of steam at 1,000 degrees Celsius. Cold water vapour was observed erupting from the dwarf planet Ceres in the asteroid belt of our own solar system, and in Jupiter's upper atmosphere. Herschel anatomized cosmic water, near and far.

It is now clear that much of the water in outer space takes the form of thin icy coatings on grains of interstellar dust. At temperatures below −170 degrees Celsius (and a lot of outer space is yet colder than that) water takes the form of ice. As the temperature warms above this level, the ice converts directly into a gas, without—because of the very low pressures in space—passing through a liquid stage. Even at cold temperatures, though, molecules of water can be dislodged from the ice layers by the energetic photons of starlight, to drift off into outer space. In outer space liquid water cannot exist: none of the water there exists as clouds in the sense we understand them, or as mist, or droplets.

Through the effect of successive generations of stars, the whole universe is chemically evolving—it is diversifying and becoming more water-rich. The clouds that are now nurseries for generations of new stars are cooler than the clouds of the early universe, and include more of the mineral grains and ice particles that make up the building blocks of new rocky planets and watery worlds. On balance, everything being equal, ocean worlds must be becoming more common as the universe gets older and grows more interesting.

The Uniqueness of Water

Most (but not all) of the ocean worlds we will discuss are water-based. This is not just because water is common enough to build worlds, but also because water is special. It is, as far as we know, one of the two main prerequisites for life (the other being carbon). There is some uncertainty here, because there are other ocean-forming liquids, even within our own solar system (as we will describe in Chapter 9). But they do not seem to have anything like water's unique mixture of qualities.

There is the shape of the molecule, for a start. Instead of the three component atoms being in a straight line, with the hydrogen atoms simply either side of an oxygen atom in the centre, the molecule instead is triangular in outline, with an angle of a little less than 105 degrees between the two hydrogen–oxygen bonds. This is because the molecule has to accommodate not only the two hydrogen atoms and the associated bond (a covalent one, in which electrons from the oxygen and hydrogen atoms are shared), but also the other electrons around the oxygen atom, that form two 'lone pairs' within the molecular structure. Spacing these out more or less evenly creates a structure that approximates to a tetrahedron, a pyramid with four triangular faces. The oxygen and hydrogen atoms therefore effectively lie at the apices of one of the faces of the water molecule pyramid.

Within the molecule, the oxygen atom grips the shared electrons rather more tightly than do the hydrogen atoms, and so the oxygen 'apex' of the molecule has a relative negative charge, and the side with the hydrogen atoms a positive one. This asymmetry (or polarity) of charge is one of the key factors that help to explain a number of water's literally life-giving properties. For instance, the positive end of one water molecule is attracted to the negative end of another, and this shared attraction, termed hydrogen bonding, helps explain why water is liquid over a relatively large (and a relatively low, for its size) temperature range. It is also a factor in the high surface tension of water, which gives surface water a 'skin' that organisms such as pond skaters can run across, and its strong capillary action (that makes it 'climb' up a narrow glass tube).

Crucially for life, the polarized electrical charge of a water molecule is also a key factor in water's remarkable properties as a solvent. When a substance that is made of ions of different charges is brought into contact with liquid water, the negative ends of the water molecules surround the positive ions and vice versa, forcing them apart and keeping them surrounded by the (relatively small) water molecules. Not everything dissolves in water (think of oils and fats) but many substances do, to a greater or lesser extent—not least the kind of molecules, such as proteins and sugars, that are biologically important. Water bodies, hence, are chemical cauldrons of such diversity and complexity that they can act, and have acted, as incubators of life.

Another feature of water relevant to planetary oceans is that, unlike most substances, it expands slightly when it freezes, through a quirk of the hydrogen bonding between water molecules. As a result, ice is less dense than cold water and floats on the liquid water surface.

This means that deep oceans can exist beneath layers of ice on Earth and (see Chapter 9) on other planetary bodies.

Imagine for a moment if the converse were true: oceans would fill with ice from the bottom up, and thick masses of ice would likely therefore fill most ocean basins to press down on the sea floor. Such ice would be screened from the Sun's warmth by the water above: it might melt very slowly from the bottom up, by geothermal heat, the liquid produced then trying to escape upwards through the massive bulk of ice. Perhaps life could exist in such circumstances—but there would be little chance of the likes of swarms of fish or coral reefs. Such a world would seem to be fit only for the toughest of microbes, not for the extraordinary diversity of complex life that we have beneath the waters on Earth.

Can other liquids take the place of water as crucible of and shelter for life? There are worlds in outer space (see Chapter 9) with lakes and seas of substances such as methane and ethane. These are rich in carbon (and likely there are some complex carbon compounds there too). Simple hydrocarbons can certainly act as solvents, as anyone who has worked in a chemical factory knows. But whether, on some other world, they can by themselves create a kind of life, is an open question, the answer to which currently seems to be 'probably not'. To have life, it seems that long-lived bodies of liquid water—oceans of one form or another—are necessary. And, for that, one needs the right kind of planet, at the right kind of distance from its star, to form out of the whirling clouds of interstellar gas and dust.

Birth of a Solar System

Star systems form in the clouds of gas and mineral dust—and of water too, as vapour or as ice particles—that drift through interstellar space. To make a star, a portion of the cloud needs to separate off, then collapse under its own weight. This might happen spontaneously, being seeded by random fluctuations in the drifting clouds. Or, as seems to have happened in the case of our own solar system, the pressures generated by a nearby supernova can trigger gravitational collapse. The clues to this trigger event lie within ancient meteorites that have fossil chemical traces of short-lived, highly radioactive isotopes that could only have been generated in a supernova that erupted just prior to our own solar system forming.

As the portion of cloud begins to collapse, any initial movement is converted into a slow rotation of the cloud. As the cloud continues to shrink, it begins to rotate more rapidly. The mass of material now occupies a smaller space, and therefore any original movement is speeded up, much as a skater spins more quickly when they pull their arms in tightly to their body. The compression of the gas in the core of the collapsing cloud begins to release heat, from the release of gravitational energy, and this heats up the interior part of the spinning cloud that is now flattening into a thin disc. The cloud becomes a luminous proto-sun, with a fitful output of heat and light reflecting the growing pains of a star in the making.

The densely packed, colliding atoms at the core of the young star generate temperatures—simply through the immense compression—of millions of degrees Celsius. This compressional heating in itself can 'burn' lithium, atomically transmuting it into beryllium. Eventually—usually after tens of millions of years—the inexorable rise in pressure and temperature makes hydrogen atoms begin to fuse into helium, releasing a truly gargantuan source of energy that can last for billions of years. A true star (or more prosaically, a 'main sequence star') is then born.

Starbirth is a thing of beauty and violence, with intense X-ray emissions and powerful stellar winds of outflowing atoms and ions. The outburst of energy powers the outflow of radiation and gas from the inner parts of the star system, driving them to its outer regions. Shock fronts develop as faster-moving packets of gas driven by these outflows impact upon slower-moving gas patches, causing further heating. The proximity of a just-lit star is no place for a volatile molecule such as water, or ammonia, or carbon dioxide. Hence, a snow line, perhaps a billion kilometres in diameter, is formed around the new star.[8]

Inside the snow line it is too hot for volatile molecules to condense, and they remain as gas—a gas that is driven outwards by the fierce solar wind. It is driven outwards until the temperatures fall so low, somewhere around the present orbit of Jupiter, that it can condense into tiny ice crystals. These can collide and aggregate into large masses of ice as they whirl around the infant Sun. It is a factory where comets are made. Water can build greater things than comets, though.

Water, here, is not simply an icy backdrop to the larger drama of planet formation. It is the motor that drives the manufacture of the largest planets of all: the gas giants of Jupiter and Saturn, and the ice giants of Uranus and Neptune. Given that water is a combination of the commonest element in the universe, hydrogen, and the third commonest, oxygen, it is little wonder that it is the commonest molecule, after molecular hydrogen, in these regions. It likely exceeds the amount of metals and silicate minerals in the circumstellar disc.

The scavenging of ice by a growing Jupiter, just in the zone where ice was condensing out of vapour, increased its mass, helping it become massive enough to attract and trap the hydrogen and helium still present in the swirling cloud.[9] In this way, Jupiter grew enormous, as did—although not to quite the same extent—its neighbour Saturn. Farther out from the new Sun's snow line, Uranus and Neptune did not grow large enough to pull in such large amounts of hydrogen. Nevertheless, they accumulated enough water to become 'ice giants'—planets largely made of rock and ice. Oceans, albeit largely hidden ones (see Chapter 9), developed in these regions too.

Planet-making in the Dry Zone

Inside the snow line, and near to the young Sun, is the zone where less volatile minerals condense out: the zone of rock and metal, where the Earth orbits.[10] There is now a race against time. The raw dusty materials that build the planets are being swept outwards even as, whirling and jostling, they aggregate into larger bodies, less susceptible to the gusts of the solar wind. It is a chaotic process—but one that is clearly efficient (for we are here on Earth to ponder it). Smaller and slower-moving[11] particles could simply 'stick' together in loose aggregates, while larger ones could smash into each other and break up. Through all of this process of building and destruction, though, dust grew into larger aggregates that in turn clumped together into planetesimals some few hundred metres to a few kilometres across, and these by collision grew rapidly into 'planetary embryos' perhaps as big as Mars. The final encounters between these produced the rocky planets that we know today. Most of this process probably took place in only a few million years.

The materials that condensed out of the fast-clearing disc of gas and dust give some clue to the kind of temperatures at which this happened. At very high temperatures (above 1,600 degrees Celsius) minerals rich in elements such as aluminium, titanium, and zirconium can condense into particles from the hot gas; particles of such high-temperature minerals are indeed found in the oldest meteorites known. As temperatures drop, farther from the young Sun, minerals including silicon and phosphorus condense. Farther out still, there are those with arsenic, copper, and silver—then, with elements such as sodium and potassium, then lead and zinc. Finally, there are the most volatile elements: carbon, nitrogen, and hydrogen. When the last of these is combined with oxygen, the resultant water vapour condenses into ice at the solar system's 'snow line', at temperatures—because of the near-vacuum pressure—well below zero degrees Celsius.

It is a kind of temperature ladder. How high up on it a planet is gives some idea of the ambient conditions, when the main phase of planet building was over, once the planetary nebula had cleared.[12] The Earth seems, overall, depleted in elements on the lower rungs of the ladder. This depletion is relative to what scientists take to be an average of the solar system's building blocks. A measure of such an average can be found in the composition of certain types of meteorites, particularly those termed chondrites, which appear to represent the stuck-together dust of the primitive nebula. Another measure is in the composition of the Sun, where the concentration of the elements (other than hydrogen and helium) can be measured by spectroscopic analysis of its bright photosphere.

These measures, when compared with estimates of the Earth's bulk composition, suggest that our planet has relatively less of such elements as potassium and zinc than the solar system average. This, in turn, suggests that, while the bulk of the Earth was forming amid the clash of planetesimals and embryo planets, these elements were still in a hot gas phase, at temperatures of several hundred degrees Celsius, and being swept outwards by the solar wind. If the primeval Earth could have difficulty in holding on to zinc and potassium, it is hard to see how it could have retained much water—a much more volatile material—as it formed.

There is some evidence, therefore (and yes, it is controversial), that the proto-Earth was dry. How, then, did it acquire its oceans, to become a blue planet?

2

Ocean Origins on Earth

At the beginning of all things, there was Father Sky, whose name was Uranus, and Mother Earth, who was called Gaia. Uranus lay over Gaia and fathered many children. These children included the Titans, and the Hekatonkheires—giants with 100 arms—and also the mighty one-eyed Cyclops. This being the mythology of the Ancient Greeks, the family did not run smoothly, and Uranus was not an ideal parent. We will, in these gentler times, draw a veil over quite how the children were treated, and exactly in what way they obtained their revenge over their father. One of the children though, a Titan, amid all the passions and the bloodiness, held aloof. He neither acted out vengeance then, nor later joined his siblings in the Titanomachy when they challenged, in battle, the gods on Olympus.

This was Oceanus, who was the god, or personification, of the world-encircling ocean. With his wife, Tethys, he simply made the world bloom and prosper, by governing the distribution of the world's springs and rivers and lakes and rainclouds. The Ancient Greeks, here, showed a nice premonition of the true nature of the global water cycle. In deriving this planet's oceans from a union of the Earth and the heavens, they may have guessed correctly, too. The truth is, we don't know. The origins of the world's oceans are still a tantalizing mystery.

The Earth's First Water Supply

The Earth, as a planet, began its existence well within the 'snow line' (and lies inside it still). At the Sun's tempestuous beginning, that snow line would have lain even farther out. The Earth, though, has oceans—and has had them for at least 4 billion years. Where, then, did the water come from, in a seemingly 'dry' part of the solar system?

First it needs to be emphasized that the Earth does not have all that much water, in planetary terms. That may seem surprising at first glance. After all, the oceans cover two-thirds of what, seen from outer space, appears as a blue planet, and a dazzlingly beautiful one. And the oceans to us are deep, averaging over 4 kilometres in depth across the ocean floors. On a human scale—and certainly to the Ancient Greeks, setting out into unknown regions in their frail boats—that is enormous. But on the scale of the Earth the oceans are merely a thin veneer. Relative to the whole planet, the depth of the ocean is thinner than the skin on an apple.

One might, for instance, take all of the Earth's oceans (and rivers and lakes and icecaps and groundwater supplies too) and roll them into a ball. The artist Adam Nieman has made a magnificent illustration of such a thought experiment, in painting what appears to be a small translucent marble rolling across the globe of the Earth. The 'marble' of the Earth's surface water is a little less than 1.400 kilometres in diameter (by comparison, the UK is about 1,000 kilometres in length). The Earth's precious cache of stored surface water, viewed thus, suddenly looks much smaller and more vulnerable. Relative to the mass of the Earth, the oceans' bulk is yet smaller, making up much less than a tenth of 1 per cent of our planet. The problem of acquiring such a small fraction of water now appears a little less formidable. But one has to think, too, of the oceans below the Earth's crust.

Something like another ocean's worth of water (and, in some estimates, perhaps as much as 25 oceans[13]) lie far below our feet, far below even the groundwater reserves into which we drill down to obtain much of our fresh water. These are not oceans, alas, like the ones complete with storms, geysers, ichthyosaurs, and plesiosaurs that Jules Verne's intrepid explorers sailed across in that nineteenth-century classic *Journey to the Centre of the Earth*. The Earth's inner oceans are dissolved in the rock and magma of the Earth itself. A tiny fragment of the mantle mineral ringwoodite (previously only known from ultra-high-pressure laboratory experiments) was recently found within a diamond in Brazil, having somehow survived its 500-kilometre journey to the Earth's surface.[14] It contained about 1.5 per cent water, lending weight to the idea that parts of the mantle represent planetary water stores.

The problem of the Earth's water is considerably larger, therefore, than that of simply explaining the surface oceans. It is a conundrum: one that is sharpened by the explosive climax to the Earth's construction.

The Late Catastrophe

There is a wild card here, which surely had an impact—quite literally—on what kind of oceans we now possess, and on the kind of oceans we might once have had, long ago. It seems almost certain now that the Earth, in its early history (some time in the first few tens of millions of years), collided with a Mars-sized planetoid, a body that has been called Theia. The frightful collision splashed out incandescent magma, which then coalesced in orbit around the Earth to form the Moon.

Such a collision would have been an extraordinarily large, and late, reprise of the kinds of collisions that had built up the Earth shortly before. The collision with Theia effectively explains the dynamics of the orbiting, spinning Earth–Moon system. However, evidence has been building that it does not explain everything. Computer models clearly show that if a Mars-sized body impacts on the Earth with the kind of glancing blow that would have given us our spin, then most of the splash material that goes on to become the Moon should come from the *impactor*. The trouble is that the chemistry of the rocks of the Earth and Moon are uncannily similar, as regards fundamental features such as the ratios of isotopes of oxygen, silicon, tungsten, and titanium; those of Mars we know to be very different (as regards oxygen isotopes, for example, Mars and Earth rocks differ by a factor of about 50).

This problem is now exercising the minds of quite a few scientists.[15] Variations on a theme have been suggested. One new model is an impact on Earth that had already been spinning rapidly because of a previous impact, and so, torn apart, could fling out a mass of its own material. Another postulates a clash of equals that are destroyed and then recombined into the Earth and Moon. Yet another idea is to have, post-collision, a long-lasting cloud of hot vapour in which the chemistry from the two bodies would mix, before the final condensation into refashioned Earth and brand-new Moon.

By whichever scenario, this was an impact of almost unimaginable violence. The material that coalesced to form the Moon was initially incandescent rock vapour and spray, from which most volatile components (such as water) would have vapourized and been driven off, ultimately to get carried away by the solar wind. The Moon is now, for most practical intents and purposes, bone-dry, both in its lack of significant surface water, and in the pristine state of the minerals within samples of Moon rock. Over billions of years, these minerals have never been chemically weathered, suggesting an absence of water.[16]

Afterwards the Earth, too, was a glowing magma ocean, perhaps to a depth of 1,000 kilometres or more. It had been refashioned to the extent that the period of Earth history before Theia has been suggested as a new eon, christened the Chaotian,[17] while the pre-impact Earth itself, being a different planet, was also proposed a new name, of Tellus. If the planet that was Tellus—a now unknowable planet—had possessed any significant volatiles (including water), then most of these should have been vapourized and dispersed into space in the aftermath of the giant impact. Therefore, if we start with an Earth constructed from metals and silicates at relatively high surrounding temperatures, and then smash another planet into it to remove what little water might have been originally there—how did we get oceans?

One answer is—later.

From the Heavens

There are other things that are strange about the Earth's crust. Something has happened to its content of elements such as lead, indium, and xenon, which are both volatile *and* have radioactive isotopes that makes them function as atomic clocks. With all of these elements, the 'clocks' seem to have been reset about 100 million years after the Earth formed.

The French geochemist Francis Albarède considers that these elements came from outer space, after the Earth had already formed. He has proposed that the Earth, both in its primordial state and just after the Theia impact, was essentially dry. The Earth's water, he suggests, came subsequently from the skies after the Moon-forming event, together with most of the Earth's crustal lead, indium, and xenon. Other volatiles such as carbon and nitrogen likely hitched a ride to Earth with these elements too, creating a 'late veneer' upon the Earth's surface. The idea is not new, but was restated persuasively.[18]

What, then, were the delivery capsules? Albarède's chosen vehicle is a particular kind of meteorite, a carbonaceous chondrite, that is thought to most closely resemble in composition the cloud of dust of our solar nebula. These meteorites formed far enough away from the Sun (in the asteroid belt and beyond) not to have been affected by the fierce heat that would have blown away most of the volatile elements from around the Earth's orbit, at and beyond the snow line.

Carbonaceous chondrites contain a lot of water-bearing mineral matter, and also various carbon compounds, as the name suggests. If the Earth had just been made of these meteorites, they would have contained enough water to supply hundreds of Earth-sized oceans (and so the Earth cannot just be a mass of carbonaceous chondrites). Much of the water within them is bound up in hydrated minerals, formed by reaction of a 'dry' mineral with water. Hydrated silicate minerals of this kind include chlorite and montmorillonite, which can be found in muds on Earth today, and serpentine, which forms when basalt minerals react with water.[19]

With chondrites, though, whether carbonaceous or not, there is still the question of how and where the water reacted with the silicates to produce the hydrated silicates. It seems that many of these reactions must take place at or beyond the snow line. There, ice can coagulate around rock, be melted by the fierce, short-lived radioactivity (inherited from that recent supernova) within the silicate minerals, and then react to hydrate those minerals.

The water-bearing rock debris must then be shifted to a closer orbit, into a collision course with the young Earth. Such orbital disturbances were probably commonplace in this early phase, as the new planets, including the giant ones, altered the gravitational dynamics of the solar system, sending debris looping in different directions.

Albarède's vision is controversial, and has been criticized.[20] The lead isotopes of that late-reading atomic clock could have been reset by much of the Earth's lead content sinking into the core, not least after the impact with Theia. Further, if carbonaceous chondrites provided most of the water, then, given the composition of carbonaceous chondrites, there should be a greater quantity of elements such as carbon and sulphur—and gold—in the Earth's crust. Rather, the critics argue, much of the water could have been smuggled in from the original, 'normal', chondritic planetesimal building blocks of Earth (with some of these, perhaps, originating farther from the Sun).

In this alternative view, the Earth might have accumulated some water as it formed,[21] held on to it despite high ambient temperatures and the Theia impact, and then later expelled it on to the surface as steam that came with volcanic eruptions. The steam then cooled and condensed as rain. Over hundreds of millions of years that rainfall could accumulate as pools, then lakes, then seas and oceans.

These are fast-moving days in this area of science.[22] One source of water that Albarède discounted—and that might help solve some of the inconsistencies—is now coming back into contention. Falling comets might, it now seems, have helped fill the Earth's ocean basins.

The Comet Connection

Comets, by and large, have had a bad press from the human race—a notoriously superstitious species—both in the past and in more recent times. The misfortune seems frequently aimed at the high-born, which might please those of more socialist disposition. Shakespeare, for instance, has Julius Caesar's wife observe, on the morning of his assassination, that 'When beggars die there are no comets seen / The heavens themselves blaze forth the death of princes'. Edgar Allen Poe (not one of life's optimists) spread the misery more widely: in *The Conversation of Eiros and Charmion*, a comet stole the Earth's nitrogen, to cause an oxygen glut and the world's end through global conflagration.

H. G. Wells was one of the few who imagined things differently. In his novel *In the Days of the Comet*, humanity was transformed by the passing of a comet. As it passed, the air's nitrogen was not destroyed but was changed into a breathable gas that promoted well-being and goodwill to one's fellow humans. This led to a utopian society in which there was—a little daringly for the time—free love, to help ease body and soul. Such delights may be beyond the powers of even extra-terrestrial geochemistry to achieve, but if comets were a key factor in bringing water to the Earth, then Wells was right to paint them as the source of life and love, rather than of doom and despair.

There has been a problem with seeing comets as ocean providers, though, that relates to the fundamental mechanics of how the solar system evolved. That comets are now coming into consideration suggests that there is something important about those mechanics that we have not understood properly.

This problem relates to how the chemistry of the Earth's water tallies with what we know of water elsewhere in the solar system. The hydrogen and oxygen that make up water are both made up of different isotopes—that is, forms of those elements in which the number of protons is constant but the number of neutrons differs. Hydrogen mostly comprises a single proton and an electron, but there is also the hydrogen isotope known as deuterium, in which the proton in the nucleus is accompanied by a neutron, to give the atom double the mass (there is also another isotope, tritium, with two neutrons, but this isotope is radioactive, decaying over time into an isotope of helium[23]). Most oxygen is made up of ^{16}O (with eight protons and eight neutrons in the nucleus), but there are also small but variable amounts of ^{17}O (with nine neutrons) and ^{18}O (with ten neutrons).[24]

When water vapour condenses to form ice, the temperature at which this happens helps govern which isotopes are brought out of the gas phase. As the temperature drops, the resulting ice tends to have higher proportions of deuterium relative to normal hydrogen. So, ice that formed in the outer reaches of the solar system can be predicted to be more deuterium-rich than that (like the Earth's water) which condensed closer in.

In recent years, humans have moved in closer to comets with spacecraft to collect particles of comet-tail, or to analyse the light that shines from them. Memorably, the NASA *Deep Impact* mission in 2005 sent a probe to crash directly into the 9P/Tempel comet, a camera on the main craft catching the sudden glare of light upon impact (Fig. 2). The first half-dozen comets to be analysed in this way turned out to have the high deuterium levels predicted—and so could not represent a major source of the Earth's water.

Recently, this simple picture has changed. In 2006, the *Stardust* spacecraft came back to Earth, having flown through the coma of the Wild 2 comet and captured up to a million tiny particles of comet dust on collectors made of aerogel, a highly porous but tough silica-based material. Analysis of the particles brought a large surprise, for there were melt-rock droplets there, and minerals that had condensed

FIG. 2. Watery comet 9P/Tempel at its point of collision with the impact probe of the spacecraft *Deep Impact*.

at very high temperatures. Thus, not all cometary material was born in the cold outer reaches of the solar system: at least some included minerals made close to the Sun and then flung outwards. This suggested a very dynamic—and therefore physically and chemically complex—early solar system.

Then, in late 2010, the Herschel Space Observatory (see Chapter 1) turned its keen gaze on to a passing comet, named 103p/Hartley 2 (Fig. 3), and obtained, as its researchers put it, 'exquisite spectra' of the

FIG. 3. Comet Hartley 2, captured by NASA's *EPOXI* mission between 3 and 4 November 2010. The water content of this comet has a deuterium value very similar to that of the Earth's oceans, showing that comets could represent a potential source of the water on Earth.

light that shone from it. These showed that the water content of this comet has a deuterium value very similar to that of the Earth's oceans—and so could represent a potential source.[25]

There is, therefore, a population of comets out there with the right kind of chemistry to have supplied at least part of the world's water. Thus, the late-ocean model for the Earth is viable once more. If there is truth in it, then there are implications. These extend not just to such matters as surface hydrology and the incubation of life. There may be revolution, too, in the workings of the whole planet.

The Hydration of Earth

Time travelling to prehistoric Earth would be a fine thing. Of course, in general it *is* a fine thing, and geologists do it every day as they walk across strata that are dried and compressed sea floors and riverbeds, beaches and deltas, swamps and forests. But for most of the first billion years of history such things are absent, long erased from the Earth's memory banks. Here, a latter-day Tardis really would come in handy, for this first billion years set the course of our planet's history, and we still know terribly little about it.

It has been called the Hadean, this early eon of Earth, after Hades, brother of Zeus and Poseidon, and the fierce and unyielding (although not unjust) god of the dead and of the underworld. To enter Hades' realm, one had to cross the Acheron, the river of pain (in later versions it is the Styx). In our modern understanding, too, water is the key to the underworld. Without water, that underworld cannot sculpt our surface world. Therefore, once water was introduced to the Earth's surface, it must descend, too, into the depths of the Earth, so it can do its work.

How much do we know directly about the Hadean, and in particular, about those crucial early years in the aftermath of the impact of Theia, and of the arrival (if such really happened) of the late water-rich veneer? The direct evidence is almost pitifully slight. It mostly comprises a few zircon crystals, each about a tenth of a millimetre across, which were once grains washed into a 3 billion-year-old sandstone that can now be found in the Jack Hills in Australia.

Zircons (crystals of the mineral zirconium silicate) are marvellous things. When they crystallize (usually deep in the Earth's crust, in a magma chamber or in the roots of a mountain belt) they can take in quite a lot of uranium—up to a few per cent—but no lead. Then, the uranium, being radioactive, breaks down into lead, which is retained within the molecular structure of the zircon. Zircons are also very tough, so they can retain these delicate atomic patterns over billions of years. Find that zircon crystal today, and carefully measure the proportion of uranium to lead using a mass spectrometer, then if one knows just how quickly uranium decays into lead one can work out the time since that zircon crystallized.

In the case of a few of the Jack Hills zircons, the ages revealed were enormous—over 4 billion years old. These did not form in the sandstone. Rather, they were washed into the sand that was to later become the sandstone from the erosion of some yet more ancient rocks—rocks that have not yet been found, and that have probably long ago been destroyed. One or two of the zircons revealed ages as old as 4.4 billion years.[26] What do they tell us, other than the mere fact of their age?

A surprising amount, it turns out. One pattern in their chemistry can suggest whether there was significant water in the vicinity when they crystallized. This is the proportion of the different isotopes of oxygen—especially that of the heavy isotope ^{18}O to the normal 'light' isotope ^{16}O. This ratio is known to be little affected by melting processes in the mantle, but to be changed in rocks that have been altered in the presence of water. The Jack Hills zircons show such changes in their oxygen isotope pattern, and thus when they crystallized water is likely to have been present *somewhere*. It doesn't mean that there were necessarily oceans at the surface, although conditions may have allowed them to exist,[27] but it does suggest that, kilometres below the surface, water was circulating. At this stage, such a finding may be even more significant than pools of surface water, for it is a hint that the engine room of this planet was becoming primed to function.

The Engines of the Earth

In a world without water things work differently—and with more difficulty—than on a world with oceans. The hot, high-pressure iron and magnesium silicate minerals at hundreds of kilometres depth in a dry planet's interior are stiff and unyielding, and cannot easily give way, even as heat builds up and those minerals begin to melt. Eventually magma rushes out on to the surface in great floods, after which release of heat and pressure the interior stagnates again, until the next great outburst. That currently dry planet, Venus (see Chapter 9), seems to behave in this way. This kind of volcanism is not good for any organisms that might try to get a toehold on such a world.

Mix in just a little water, though, and the minerals soften and weaken. These rocks, deep down, can therefore flow more easily. Crucially, if magma comes up in one place causing the Earth's rigid lithosphere (the crust together with the uppermost part of the mantle) to stretch and break apart, then other sections of crust can begin to founder and sink through into the mantle. As these sink, they pull on the adjacent lithosphere, and that pulls apart yet farther, allowing more magma to the surface.

This is the beginning of plate tectonics, in which the plates are separate and independently moving sections of the Earth's lithosphere. In Francis Albarède's view, the water that arrived in a shower of water-bearing meteorites (and, as we now suspect, comets) began, over many millions of years, to get mixed in with the crust and upper mantle. At some point, sufficient water had been added to allow the Earth's plate tectonic motor to begin to run. Once triggered, this motor has been running ever since, smoothing and modulating the processes of volcanism and earthquakes, and creating the Earth's mountain chains and ocean basins. It set the scene, *quite* certainly, for the evolution and maintenance of life.

Plate tectonics, for all the periodic death and destructiveness caused by the great volcanoes, earthquakes, and tsunamis associated with its working, is still a remarkably gentle process compared with all the alternatives. In essence, it allows the Earth's heat to be released smoothly and evenly, and it recycles and renews much of the Earth's surface, piece by piece.

The 'folding in', as Albarède put it, of the Earth's surface water deep into the interior is something of a mystery process, for there is little direct evidence of that 'start-up' phase of plate tectonics—although that has not stopped people thinking about how this might have happened (see the section 'The Last Great Readjustment'). Once started though, plate tectonics would certainly have been helped by the waters of the early oceans that literally acted as a lubricant (as they still do) to help ease the path for descending crustal plates at subduction zones.

As regards the oceans, however, the beginnings were likely anything but smooth. For most of the next billion years, rocks were still raining out of the sky.

Perilous Times for the Hadean Oceans

The 'late veneer' that arguably modified the Earth's crust and brought in the world's water stores was not the end of the meteorite bombardment. It takes quite a while to clear a young star system of its debris, or to arrange the remains of that debris—as in the asteroid belt—in more or less stable orbits. And so, over the millions of years that followed, high numbers of large asteroids and comets continued to fall out of the sky and impact upon the Earth. (Some still do, of course, although much more rarely now.)

This planet has lost almost all evidence of those early violent events. The continual renewing of the Earth's surface, thanks to plate tectonics, has erased those enormous early craters. Evidence of that bombardment can be seen on the face of the Moon, where the absence of air, water, and plate tectonics has left the impact craters in almost pristine condition. All that has modified them are more recent impacts and, in places, the basalt lava that has oozed out to form the dark 'seas' that partly cover them. The surface is so heavily cratered that it is difficult to extract a clear history from it (Fig. 4). However, the Moon has recently been mapped in great detail using lasers by an orbiting NASA instrument rather fetchingly called 'LOLA' (for Lunar Orbiter Laser Altimeter). The results showed that the ancient highlands were hit by a range of meteorites, from small to very large, while the meteorites that impacted the main basalt 'mare' (that are about 3.6 billion years old) were small to medium sized.

The Earth must have been yet more fiercely pelted, being both a larger target and having a much stronger gravitational field to draw in meteorites. Several times over the first half-billion years of the Earth's history there must have been meteorite impacts large enough to vapourize all of the Earth's oceans, and sterilize any life that had managed to evolve within them. Could anything survive at all? Perhaps—even if all the oceans did repeatedly boil away. Today, microbes extend down into a 'deep biosphere', inhabiting tiny rock fractures to depths of a kilometre or more. If that was the case in the Hadean, then maybe there could be survivors (at least of the less cataclysmic of the large impacts), to emerge to recolonize the newly condensed oceans, as the Earth cooled over the millennia following a major impact.[28]

FIG. 4. Craters on the eastern limb and far side of the Moon, imaged by the *Apollo 16* spacecraft in 1972.

There was, it has been suggested, one last such event that has been called the Late Heavy Bombardment, which is thought to have taken place some 3.9 billion years ago as the orbits of Jupiter and Saturn readjusted, perturbing the orbits of many of the asteroids near to them and sending them shooting across the solar system. A model for such a profound reordering in the solar system—a far cry from early ideas of stable planetary orbits—was proposed by scientists working in Nice, France, in 2007 (hence it is now know as the 'Nice model'). This was the last great clearing of the solar system, and, once it passed, the oceans of the Earth were no longer under the threat of sudden vapourization. Other threats were to emerge in the succeeding eons, but the oceans could now adjust themselves to the contours of an evolving Earth; an Earth that might have been preparing to evolve *radically*.

The Last Great Readjustment

How did plate tectonics start—and from what? With little scope for an answer on Earth, one has to look farther afield for clues. Fundamentally, plate tectonics is a means of releasing planetary heat. If a planet is small and cold—like the Moon today—simple conduction of heat through the rock mass will suffice. But with a hotter body that won't do, because rock is a poor conductor of heat. So the heat builds up until the rock melts, and the molten rock makes its way to the surface where it can radiate its heat out to space. The question then becomes *how* does the magma travel.

The Earth in its early days would have been much hotter than now, with a lot of heat still retained from the energy of the impacts that constructed it, and with more internal radioactivity. A comparison has been drawn, remarkably, with one of Jupiter's moons, Io. This might be far from the Sun but, stretched and squeezed by the immense tidal forces exerted on it by its giant parent planet, its interior is heated so much that the flux of heat from interior to surface is about 40 times that of the modern Earth. This heat is released as magma outpourings on to the surface, and Io is by some way our solar system's most active planet. There is no evidence that its volcanism is linked with any form of super-energetic plate tectonics. Rather, the magma must find its way to the surface through simple vertical conduits—or, as they have been called, heat pipes.

The hot early Earth, prior to plate tectonics starting up, was essentially a heat-pipe planet too, according to the US geologists

William Moore and Alexander Webb.[29] Its thick, one-piece crustal shell could have been so effectively traversed by the heat-carrying pipes that it might have remained relatively cool at depth (something that has been inferred about the earliest crust from the oldest Precambrian rocks, and that, before the heat-pipe idea, had always been a puzzle). The magma would pour out on to the surface at, on average, about a millimetre a year—and so slowly, incrementally, build up its thickness.

Where can one find rocks from heat-pipe Earth? Some of the oldest rocks on Earth, the 'greenstone belts' of Pilbara in Australia and Barberton in southern Africa, may have formed on such an Earth more than 3.5 billion years ago according to Moore and Webb. They show kilometres-thick outpourings of basaltic rocks (presumably out of the heat pipes), separated by sedimentary strata eroded from these volcanic rocks. Some of the volcanic rocks tell of an Earth that was then hotter, deep down—dark, dense rocks called komatiite. These are related to basalts, but possess even higher amounts of iron and magnesium and so were erupted at temperatures a couple of hundred degrees Celsius higher than typical modern basalts. They are testimony to the hotter inner Earth of the past.

The Moore and Webb interpretation is a controversial one. Most reconstructions of the early Earth invoke some kind of non-standard, fast-running early variant of plate tectonics. However, if the heat-pipe model is true it suggests a very different topography to Earth—and therefore a very different style of creating low areas (where lakes, seas, and oceans might pond) and high ground (the landmasses).

On such an Earth, what might build the topography of land and sea? One phenomenon known from these terrains is the domes made by masses of rising magma pushing upwards on the crust—with the adjacent ground sagging in between. If this process controls topography, then the whole pattern of ocean/continent geography—and the fundamental proportions of land area and ocean area—would have been utterly different to today's familiar patterns. An alien planet on our own Earth indeed.

But what do the rocks tell us about the nature of the sea on the early Earth? From 3.8 billion years ago we find preserved rock strata, from which one can say that there was, then, surface water. The earliest rocks of all, terribly mangled (but still partly decipherable) rocks from Greenland, are made up of particles of former sediment—sand and mud—that could only have been laid down in water, although whether the water bodies were lakes or seas, shallow or deep, is unclear.

A little later, there is some evidence that the early Earth may have possessed *deep* water. In the 3.5 billion-year-old rocks of the Pilbara area of Australia, there is a particular combination of strata, including what were fine, silica-rich muds, pebble layers, and slumped, contorted deposits that have been interpreted as deep-sea floor deposits—the kind of sediments that pile up on the edge of an ocean floor today.[30]

A true ocean? Perhaps. It is the best evidence yet found, but it is by no means conclusive, for rock patterns of that kind simply indicate that there has been transport of sediment from shallow to deeper water, without specifying how shallow is shallow, and how deep is deep. There is work to do still, to decipher the shape and depth of those ancient seas.

When would the plate tectonics have started to usher in the kind of planet, and the kind of ocean basins, that we know today? A number of curious changes seem to have taken place around 3.2 billion years ago, still in the depths of the Archaean Eon. Changes occur in zircon crystals, and in the chemistry of the element hafnium[31] that suggest a change in the rate at which the Earth's crust was being formed. The first undoubted traces of modern-style mountain building likewise appear at this time, and change in the world's diamonds also took place. It's a curious story—but very revealing of just how mysterious are the ways in which the Earth moves.

Deep down in the mantle, diamonds are now forming, and have been forming since the Earth began. These highly sought-after crystalline cages of carbon need very high pressures to form. On Earth those kinds of pressures start at around 120 kilometres below the surface. These diamonds must then reach the surface by high-speed transfer (otherwise they would alter to graphite on the way up). This transfer is provided by a very special kind of volcano, one that is as rare as the diamonds themselves. It is called a kimberlite, and kimberlite eruptions are so rare that none have been witnessed by humans. Kimberlite eruptions are deep-rooted and extraordinarily gas-rich, and, just now and then, over the aeons have punched neat cylindrical holes—kimberlite pipes—in all of the world's ancient continents, filling them with minerals (including diamonds) derived from hundreds of kilometres below the surface.

Not all diamonds are pure and flawless. Many have tiny inclusions of minerals and rocks that crystallized at those great depths in the mantle. Trapped and preserved for billions of years within that hardest of adamantine shells, those mineral inclusions, just a few thousandths of a millimetre across, can now be analysed, routinely, by the modern magic of scanning electron microscopes and electron microprobes. The story they tell is more valuable than the diamonds themselves.

Analysis of many diamonds from around the world[32] has shown that those which are up to 3 billion years old commonly contain tiny particles of a rock called peridotite—which is basically mantle rock from which the stuff of ocean crust has been extracted, and also of a rock called eclogite, which is what subducted ocean crust turns into once it has slid down to 100 kilometres or so below the Earth's surface.

This represents good evidence that, as far back as 3 billion years ago, ocean crust was forming, and was also being destroyed by being subducted to great depths in the mantle. That is a modern-like Earth in action.

Diamond inclusions a little more than 3 billion years old, though, are different. They still contain peridotite, but there are none of eclogite. It is a tiny and seemingly insignificant change in mineralogy—but the implications are momentous. It suggests that while the ocean crust was forming on a very ancient Earth, deep subduction was not taking place. Plate tectonics, therefore, in the sense we understand it, appears to have started up on Earth just over 3 billion years ago.

The original explanations for this puzzling change in diamond composition was that early Earth tectonics could have been carried out via mantle plumes.[33] These are enormous, slow-moving fountains of mantle material that move vertically up over hundreds (or perhaps thousands) of kilometres. Where they reach to just below the Earth's crust, they stimulate increased magma production and outbursts of volcanism. Mantle plumes today are a controversial topic, but there is good evidence for them beneath places like Hawaii (the chain of islands forming as the ocean crust moves slowly over a stationary plume top) and Iceland.

Strong plume activity in the early Earth could have produced volcanic centres that—perhaps—could have seeded the first continental cores. There also were likely mid-ocean ridges, formed as a result of the hot mantle convecting upwards. If that was the case, then there must have been some kind of downwelling mechanism too, to counterbalance it. That has been envisaged as producing a kind of stacking ('imbrication') of crustal slices above the point where the mantle currents turn downwards, but without deep subduction of that crustal material. Melting associated with those stacked crust piles would lead to further production of early crust. For 'mantle plumes' one might, in Moore and Webb's parlance, now say 'heat pipes'. These are also vertical, cylindrical conduits for magma ascent. By whichever name, this kind of 'precursor tectonics' might have been more effective at producing continental crust than the plate tectonics that seemingly succeeded it some 3 billion years ago. Well before the heat-pipe model was dreamed up, one commonly cited pattern was that continents grew rapidly until about 3 billion years ago, and then the rate of their growth slowed. Perhaps that, given this new idea, is not a coincidence.

What caused the change to a plate tectonic Earth? It is tempting to link this with the arrival of water on Earth via comets, and its infolding into the depths of the Earth. If such infolding of material (both rock and water) into the mantle happened on a heat-pipe Earth, this might have increased the water content of the mantle to a point at which 'modern' plate tectonics started. Or perhaps the transition was simply caused by the cooling of the Earth to a threshold where cold ocean crustal slabs could sink. It is still very early days for the science of deciphering the early Earth.

The hypothesized[34] switchover from heat-pipe Earth to plate tectonic Earth has been envisaged by Moore and Webb as rapid—a flip from one planetary state to another. With a new mechanism of heat loss—pouring magma out on to the surface from ever-open cracks, thousands of kilometres long, the heat pipes would have congealed and died. Perhaps, though, mantle plumes might be regarded as the remaining few. Or perhaps heat pipes will return as shadows of their former grandeur towards the end of our Earth (see Chapter 8). For now though, we can take up the extraordinary story of our familiar Earth, and of how it creates ever-changing shapes in which oceans can be held.

3

Ocean Forms

Isn't it lucky that we've got just enough water. Not too much and not too little. How much, though, might be too much?

Let us imagine a proper water world—one where even the tips of the highest mountains are covered. Hollywood, of course, has got here first. In that breathless cinematic epic *Waterworld*, Kevin Costner sailed the high and endless seas of a future submerged Earth, global warming having done its stuff. He did battle with the most villainous and eyeball-rolling of pirates, while protecting the young heroine who just happened to have, tattooed on her back, a map of the world showing the whereabouts of the very last island.

Amid such improbable plot devices, and leaving aside the awkward fact that even melting all the world's ice would leave quite a lot of landmasses standing high and dry, *Waterworld* does give occasional pause for thought. For instance, on a real water world weather patterns would be substantially different. In the film, the whole action happens in bright sunshine under a cloudless blue sky. This framed the derring-do nicely, but in reality would be highly unlikely—all the energy from that evaporating water vapour has to go *somewhere*. Further, without a landscape to chemically weather it would be more difficult to control carbon dioxide levels in the atmosphere, rendering that perfect weather yet more unlikely.

There was more than weather control to contemplate in the film, though. The half-crazed human survivors, living on their makeshift raft homes, did not much go in for fishing, as short of food as they were (although our hero did single-handedly overpower the obligatory sea-monster, prior to dispensing monster steaks all round). Here the perspiring scriptwriter might have been on to something. On the real Earth of today, part of the water evaporated from the oceans later falls on to land as rain. The water, passing through rock and soil, dissolves a variety of mineral and organic nutrients, then returns that to the sea via river flow. In the sea, those washed-in nutrients feed the growth of the plankton that is the base of the rich and diverse marine biology that we know (see Plate 1).

Far from the land, in the middle of the oceans, there are what amount to ecological deserts, where primary productivity (through photosynthesis) in the sunlit surface water is very low because the nutrient flood does not reach so far. These nutrient scraps are soon used up by what few plankton there are, which die and sink into deep water. What little replenishment does arrive comes as far-travelled windblown dust (blown from distant lands that would be absent in a water world), or is stirred up from deeper water by storms.

Something like that would be the ecology of a true water world. It would not be a completely biologically sterile world. There would still be those nutrients at depth, dissolved from undersea volcanoes and such, but it would be hard to get those into the sunlit surface waters. A real water world would be impoverished compared with our current riches. On the kind of tenuous food base that would result, sea monsters would find it hard to scrape a living. It would not be easy, either, to evolve energetic, intelligent, complex mer-people on such a world.

Conversely, take a world with not very much water: one where that water would be restricted to a few scattered shallow seas and a scattering of lakes and ponds. Here there is much less scope to start or maintain a hydrological cycle, or to sustain a grand planet-wide cycle of plate tectonics. The water in such a pitiful scattering of water bodies would not be the pure and limpid water of storybook desert oases, but likely a brine so concentrated that it would suck any scraps of moisture from the air—and be a huge challenge to the workings of any biological cell that might try to survive in such conditions.

These alternative visions of the world are by no means far-fetched—as we shall see when we lift our eyes to the skies to scan the solar system and the distances beyond (Chapters 9 and 10). On Earth our present happy condition—as we perceive it, knowing little of any other—may be put down to sheer chance at work within a cosmic game of billiards. A few more or a few less large, water-bearing asteroids or comets impacting on the Earth could have made the difference between a desert Earth and a land-less water world.

In the game of cosmic chance, the Earth ended up with just enough water to irrigate the land and fertilize the seas and scrub the air of too much carbon dioxide. There has been enough water, too, to kick-start the Earth's plate tectonic machine. That last feature has been particularly convenient, because it has given the Earth the basins to hold its water in.

The Shape of the Oceans

If one was to take the solid Earth and simplify it to its absolute basics, then it is, simply, a sphere. The waters on such an Earth would be consistently some 2.5 kilometres deep. More precisely, the Earth is a sphere that is slightly flattened at the poles, because its spin is causing it to bulge out at the equator (the water would bulge slightly, too).

Let us look in just a little more detail. We have a world of two levels. Most of the world's solid surface is 4 kilometres or thereabouts below sea level, as an array of more or less flat plains (that, very slowly, slope upwards to long, low ridges that snake across this overall level). These areas are, geologically, the ocean basins and mid-ocean ridges. The ocean waters are held within these basins, but also overspill them a little on to the higher of the two levels.

This second level is about a kilometre above sea level. It is our familiar terrestrial landscape, but also extends beyond it, across the shoreline, on to the shallow sea floors that surround us. Together, the land and shallow sea make up the geological continents, which occupy a little less than a third of the Earth's surface. Between these two levels, of continental and ocean floor surfaces, there is a distinct step called the continental slope. It is not quite a cliff, but rather forms a marked slope that descends more or less steeply from the higher level to the lower one. We cannot see it directly, because it is entirely under water. It is, though, the most important boundary in the world. It separates two fundamentally different types of the Earth's crust.

Two further elements can be added. On the continents there are the raised, rugged masses of the mountain chains. These may run along the edges of the present-day continents, like the Andes, or stretch across them, like the Himalayas or the Urals. To mirror these, in the oceans there are long narrow trenches (rather narrower than are the mountain belts). These are the ocean trenches, which descend in places to quite monstrous depths, most famously in the bottom of the Marianas Trench in the western Pacific Ocean, 11 kilometres below the sea surface.

Finally, let us take just two more elements. One is the lid that keeps the water in, high in the sky. It is a cold trap at the base of the stratosphere, where temperatures fall so low that water vapour, so much a part of the underlying troposphere, freezes into ice crystals within clouds, that then fall back to lower levels. The stratosphere, hence, is dry, and has, up to now, functioned as an effective seal to keep the Earth's water in. Not all planets are so lucky (see Chapter 9), while our own luck in this respect will not last forever (see Chapter 8); still, so far, so good. Then there is the lower seal to the oceans, below the ocean crust, where the crustal minerals can no longer absorb water (that is, become hydrated) because temperatures and pressures become too high.[35]

So that is the world in a nutshell. There are a *great* many more levels of detail, but for the moment this is all we need to ponder the container that the ocean waters are held within. The most fundamental structures of the Earth—the ocean/continent divide—was, however, a late discovery even within this newest of sciences.

What, for instance, might the greatest all-round natural scientist of the early nineteenth century have understood of the oceans? Our votes for that particular title would go to Alexander von Humboldt (1769–1859), a man revered by Charles Darwin. Humboldt came closer than any person alive or dead to integrate the sciences—even as they were rapidly diversifying—to give as complete a picture of the Earth as was possible in his day. As a young man he was a noted explorer, travelling through South America and collecting data on geography, geology, biology, meteorology, languages, and ethnography (he was on the side of the angels, too, being scathing about the exploitation and slavery that he witnessed). On his travels he climbed higher than any man had before, on scaling the mighty Chimborazo in the Andes. Humboldt wrote up his South American travels in many large volumes that Darwin took with him on the *Beagle,* and quoted from in his own writings.

Late in life, Humboldt synthesized his unsurpassed understanding of the Earth in a single short[36] work, *Cosmos,* published in 1845. Amid the sophisticated discussions of everything from comets to fossils, volcanoes to climate, the few short pages on oceans amount to a brief essay in perplexity. The depths of ocean, he said, were 'unknown to us', noting that in some places sounding lines more than 6 kilometres long had failed to reach the sea floor.

Humboldt wondered, too, at the way the coastlines of South America and Africa seemed to match up, although he did not so much look forward in this to ideas of continental drift and plate tectonics, as seem to glance backwards a generation to the ideas of the Comte de Buffon (see Chapter 9), who had contemplated the world in pre-revolutionary France. Buffon had thought that the pointed shape of South America was due to the flood of the Earth's primordial waters, which condensed and fell as rain at the poles (falling there because those regions cooled more rapidly than did the tropics) then rushed towards the equator, scouring the landscape as they went. Humboldt thought that the Atlantic Ocean resembled a meandering valley, directing the flow of the eddying waters first to the west, then to the east.[37]

Humboldt saw from geological strata that the positions of land and sea could change place, and was aware that the Earth's crust could rise and fall. That did not mean that continent and ocean were completely interchangeable. He thought the continents owed much of their bulk to the 'eruption of quartzose porphyry',[38] thus considering continents essentially as large volcanic masses rooted to the sea floor.

Humboldt's perplexity concerning the fundamental structure of the planet can be forgiven. For how could one study that two-thirds of the planet covered by water when that water, in bulk, was as opaque as, and more impenetrable than, rock? At least miners could dig down through rock, while no one, then, could swim to more than a few metres below the surface of any mass of water. One could take soundings from ships, but that was easier said than done. On a ship in the nineteenth century that involved the reeling out of a rope (later designs used string twine, then woven piano wire) with a weight or bucket (to capture bottom sediment) on the end. For shallow water, that was relatively quick and simple, but for truly deep water, that needed reeling out a few *miles* of line. How did one know when that line hit the bottom? This needed great precision and skill, watching the speed of the line to be able to judge a change in the way it reeled out. The potential drag effect of deep-sea currents on the line added further uncertainty, too. All in all, such gathering of data was a slow, tedious, and uncertain business. Within even the best-known ocean, the Atlantic, long a regular shipping route, there was then no inkling that such a thing as an enormous ridge ran down its middle.

The first suspicion of this submarine topography came from a man almost as extraordinary as Humboldt himself. Matthew Fontaine Maury was an energetic, adventurous Virginia farm boy, seemingly destined for the farming life. At the age of 12, that spirit of energy and adventure led him to climb a high tree. He fell from it and hurt his spine so badly as to end the prospects of farming. Sent to school, he showed an almost unbelievable tenacity of study (one that he was to maintain throughout his life), learning Latin grammar, for instance, in seven days. Against the wishes of his father he joined the Navy, sailed the seas of the world, and not only learned the skills of navigation but determined to improve them.

Noting, for instance, that there were no systematic records in the Navy of the winds and currents to be encountered in voyages, he simply set about compiling these to produce the first wind and current charts of the Atlantic. He prepared star charts too, so precise that they were used in back-calculating the orbit of the planet Neptune, then newly recognized. He conceived the system of meteorological observation that later became the US Weather Bureau. Caught on the Southern side in the American Civil War (although he hated slavery) he—alas—invented the all too successful electrically controlled underwater mine.

It was Maury's 200 carefully organized soundings across the Atlantic (carried out to enable the first transatlantic cable to be laid) that first showed the existence of a higher region in the middle of this ocean. His 1855 map admittedly shows the ridge as a vague, elongated blob in the North Atlantic. Nevertheless, this underwater 'plateau' was the first hint of the ocean's grand underlying structure. Humboldt himself was impressed by the work, which was not only technically meticulous but evocatively described (Maury, on his long voyages, used to read and reread the works of Shakespeare). Maury had, Humboldt said, invented a new science: the physical geography of the sea.

Towards the end of the nineteenth century, the pioneering oceanographic ship HMS *Challenger* took further soundings, and confirmed the existence of the mid-Atlantic 'plateau'. It was one of many achievements of this astonishing little ship: 60 metres long and 12 metres across, squeezing in—or starting out with, more precisely—269 men.[39] Between 1872 and 1876 it sailed 100,000 kilometres across the oceans, through all climates, carrying out brutally hard and demanding work.[40] The result was the first systematic picture of the oceans: of the general shape of the ocean basins, of the nature of the waters, and of the deep-sea sediments and animals that were brought up, day after day, by the dredge buckets from 4 kilometres or more below the sea's surface. Its results, described in 50 volumes, were the beginning of oceanography as an organized science.

The true shape of the Mid-Atlantic Ridge, though, only began to be slowly revealed after the First World War, when echo-sounding—a more illuminating technique than the *Challenger*'s sounding lines—began to be used systematically to reveal the shape of the ocean floor. And people only began to *see* what the ocean floor looked like after the Second World War. A remarkable scientific partnership then began to combine art and science to conjure up perhaps the most iconic images of the oceans ever made. Matthew Maury, in one of his many vivid phrases, had once written, a touch rhetorically—'Could the waters of the Atlantic be drawn off, so as to expose to view... the very ribs of the solid earth... the cradle of the oceans?' This is exactly what Bruce Heezen and Marie Tharp set out to do in the 1950s—and in which they succeeded magnificently.

They both worked at the Lamont Geological Observatory of Columbia University in the USA. Bruce Heezen, an oceanographer, sailed in the Observatory's ship to collect data on the sea floor. Women were not then allowed on board ship, so Marie Tharp stayed behind and drew the maps made from the data that Bruce collected (she did not get to go on an expedition until 1965). The first maps were produced in 1959. Most people around the world saw them as the memorable supplements to the *National Geographic*, the Atlantic map being the first of these, published in the summer of 1968. They were quite *literally* memorable: for people of that generation had never seen the like before. One author of the words you are reading was then an impressionable 14 years of age, and vividly recalls the moment of unfolding the map.

The sea floor, once the water was removed, looked much more dramatic, much more like something from science fiction than anything on land. Even the mighty Cordillera did not compare to the Mid-Atlantic Ridge, which marched almost from pole to pole, with an extraordinary array of fractures that cut across it. No wonder that Maurice Ewing (Heezen and Tharp's boss at the Observatory) talked of 'millions of miles of a tangled jumble of massive peaks, sawtoothed ridges, earthquake-shattered cliffs'.

It is a work of art, and in many ways an *imaginative* work of art. For the map looks superbly and convincingly detailed, and yet Heezen had not been able to collect anywhere near enough data to map the sea floor in such detail. Marie Tharp, employing much intuition and a good deal of creativity, filled in the many gaps in data with the kind of patterns that she divined *should* be there. Her powers of inference were considerable. From the early data, back in the early 1950s, she began to suspect that the top of the Mid-Atlantic Ridge was not, in fact, a ridge. Rather, running right down the middle of this enormous elevated mass there was a notch: a narrow valley, that she thought resembled a rift valley, where a long central plug of rock had slid downwards between two parallel fault planes. 'Nonsense,' snorted Heezen, 'that's just girl talk.' Marie Tharp stuck to her guns—and she was right. She had discovered one of the fundamental structures in the ocean (Fig. 5). It showed that the ridge was not so much being pushed up as pulled apart (and Heezen was later to give her credit for being right). It's a pity, though, that the realization was to lead them both, for a time, up a very bizarre garden path.

The Ocean Machine

One of the reasons why Bruce Heezen didn't at first like the idea of a rift valley running the length of the Atlantic was because, like many geologists of the 1950s, he thought that continental drift was nonsense. Yes, the edges of the continents did seem to fit together in an uncanny sort of way, as Alfred Wegener and others had suggested years ago. And yes, the crust of the oceans was different, being denser (because of having more iron and magnesium and less silica) than that of the continents. This explains why the continents are raised high, and why the ocean floor lies several kilometres below them. But to plough the continents for thousands of kilometres *sideways* through the oceans? That was absurd, surely.

However, when Heezen convinced himself of the rift valley in the central Atlantic, and saw that the whole ocean must be pulling apart along the rift, with basalt lava welling up to fill this ever-opening fracture, he agreed that the oceans must be getting bigger. This needed

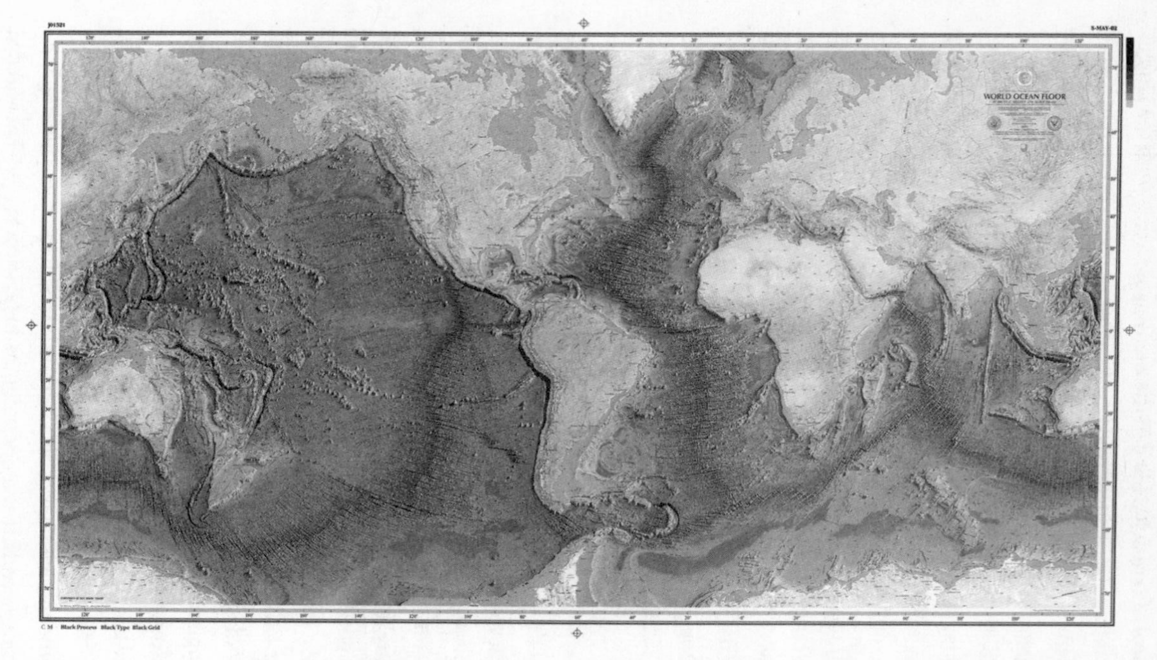

FIG. 5. The Heezen and Tharp map of the ocean basins: the mountain ranges of the oceans are evident, especially the mid-ocean ridge running the length of the Atlantic.

explaining. There was a considerable space problem here. If the Atlantic was widening, then all that crust had to go somewhere.

One of the ways to find that 'somewhere' was to increase the surface area of the whole globe—that is, for the whole Earth to expand. This idea had been expounded by Warren Carey, an Australian geologist in the 1950s. For some years both Heezen and Tharp subscribed to this idea. Where did the matter come from? Here, matters were a little vague, and Carey himself linked the expansion, a little mystically, to the expansion of the whole universe. Others thought that the power of gravity may be decreasing, causing the Earth and other planets to expand from the release of pressure. The ultimate reason, to the 'expansionists', did not seem to matter so much (after all, there are lots of things that people don't understand). What did seem to matter was the empirical fact, as they saw it, that the ocean basins were getting bigger.

In the mid-1960s, there came the scientific revolution that was plate tectonics (which Heezen and Tharp both soon joined). This explained the continual expansion of the ocean crust at the mid-ocean ridges that Heezen and Tharp had so entrancingly depicted, and the drift of continents. New evidence also came into play. Magnetic 'stripes' had been detected in the ocean crust by airborne surveys, symmetrically disposed either side of the ridge. They reflected production of ocean crust at the ridges successively through times when the Earth's magnetic field was like today's, with magnetic north at the geographic North Pole, and times when the magnetic field had flipped so that the magnetic pole was in the south (as was last the case a little under 800,000 years ago). This pattern had been noticed by the British geologist Drummond Matthews and his student Fred Vine in the early 1960s, and was a crucial part of the puzzle. The plate tectonics hypothesis squared all these circles simultaneously by continuously producing ocean crust at the ridges, and then sliding it back into the mantle at the ocean trenches. It is still the most persuasive and fruitful of hypotheses and, through the almost impossibly precise measurements of movements of the Earth's crust provided by satellite-borne lasers, it has essentially become science fact.

There are still a few small groups of 'expansionists' around the world, holding on to their vision with all the tenacity of a tiny and impassioned religious sect, and making elaborate models of a world that they think has puffed out like a beach ball being blown up for a day out at the seaside. And one has to remember that this would have happened only in the last 200 million years (for that is the maximum age of preserved ocean crust). It is still a striking, even outrageous concept—and there are converts among the young and scientifically romantic to be gained too, for the concept is now daringly anti-establishment. Alas for this beautiful idea, measurements of the Earth in the satellite age clearly show that the Earth is not expanding—or at least not expanding by more than a fifth of a millimetre a year (the current uncertainty of measurement). Plate tectonics has gone from being a revolution to humdrum normality in one human generation. That, though, is how the world works.

It is almost an equally shocking idea that the ocean plates—slabs of crust and upper mantle about 100 kilometres thick—are continuously inching their way across the Earth (almost literally inching, as measured speeds range from less than 1 centimetre to more than 10 centimetres a year). Created as hot (and therefore relatively buoyant and therefore higher) crust of the mid-ocean ridge, the ocean plates, over time, separate ever more widely, cool, become denser, and sink lower. Eventually they reach the end of the line.

Once an ocean plate reaches an ocean trench, it bends and sinks downwards back into the mantle. Sliding one gigantic slab of rock past another is not a gentle process. Part of the movement, it is now known, does take place as slow 'creep', a process that does not cause earthquakes. About half of the movement, though, occurs by sudden fits and starts, where the boundary between the two plates initially locks up through friction. Pressure then builds up until the down-going plate suddenly, catastrophically, slips, unleashing enormous amounts of energy in the form of earthquakes and tsunamis.

The ocean basins are hence young by comparison with the truly ancient and permanent continents, which may be up to 20 times (or more) older.

We now know a great deal about the oceans, because of a scientific undertaking that has been revolutionary, but that has been carried out in near obscurity and remains largely unsung. It has been mostly a hidden revolution, without fanfare and away from public gaze—and yet it has transformed our understanding of the Earth.

The Albatross and the Ocean Revolution

The history of scientific drilling in the oceans starts with what must count among the most cherished and—alas—short-lived of scientific institutions. This was the American Miscellaneous Society, or AMSOC for short. AMSOC was set up in 1952 by Carl Alexis and Gordon Lill, geophysicists at the Office of Naval Research. They had been going through a pile of scientific proposals, each of which could not be categorized as anything other than individual: hence, they were miscellaneous. From that, it was but a short step to the creation of AMSOC.

As the oceanographer Willard Bascom later recounted, any scientist could claim membership, because there were no membership rolls—although he denied the rumours that only those with research proposals too far-fetched to receive government funding could be admitted (it was, he said, merely coincidence). Nor were there bylaws,[41] officers, or membership dues. The motto of AMSOC was classically coined *Illegitimi non Carborundum*, which roughly translates as 'Don't let the bastards grind you down'.

The only formality was an official award, the Albatross Award,[42] of a real stuffed specimen 'borrowed' from a storeroom of the Scripps Museum. Recipients included the founders of the award, Gordon Lill, John Knauth, and Arthur Maxwell (in 1959, because they worked out that they might not otherwise be nominated); the oceanographer Henry Stommel in 1966 (for 'abandoning oceanography's most cherished chairs'—he had moved between several distinguished professorships in a short space of time); and the even more distinguished geophysicist Sir Edward Bullard in 1976 ('for unintelligible magnetism'—his studies had been found hard to comprehend).[43]

Amid this inspired lunacy, AMSOC was the moving spirit behind one of the most ambitious scientific quests to explore the oceans—the Mohole Project.

This was the idea to drill through into the deep ocean to reach the Moho—more fully known as the Mohorovičić Discontinuity, named after the Croatian geophysicist and meteorologist Andrija Mohorovičić, who was the first to discover it. The Moho is the interface (visible in geophysical soundings of the deep Earth) that separates crust from mantle. Beneath the continents this boundary usually lies more than 30 kilometres below ground, but ocean crust is thin and it is often less than 10 kilometres below the sea floor.

Extraordinarily, the project got underway. In 1961, five holes were drilled off Guadeloupe Island in Mexico. None of them got deeper than 200 metres below the sea floor. No matter. They had done this from a floating platform, through over 3.5 kilometres of ocean water—the first time such a feat had ever been carried out. One of the holes, too, had gone through the surface sedimentary layers into basalt—the rock that we now know all ocean floor crust is made of.

The Mohole Project did not get further, because it was clear that it was going to cost a lot of money to reach the Moho (indeed, this has still not been achieved). AMSOC wound up the structure—that it did not, of course, have—in 1964, although the much-travelled albatross continued to be awarded until at least 2002.

The Mohole Project, though, had shown that drilling through the deep ocean, from a drilling platform placed on top of a ship, was possible. From that, there formed a project that started in 1966 as the Deep Sea Drilling Project, was renamed in 1985 as the Ocean Drilling Project, and subsequently mutated in 2003 into the Integrated Ocean Drilling Program, which in 2013 underwent a further change to the International Ocean Discovery Program.

No matter. It has been one of the great, and largely unsung, revolutions in the Earth sciences. In just the first phase, the Deep Sea Drilling Program, the ship *Glomar Challenger* sailed over half a million kilometres from the tropics to polar regions, and drilled over a thousand boreholes totalling 170,000 metres in up to seven kilometres of water.[44] The many shipboard scientists who have taken part would work 12-hour shifts, examining the core for rock type, mineralogy, physical properties, chemistry, and fossil content. The results went on to form an impressively weighty collection of blue-bound volumes, and the science that came from it rewrote the history of the world. Most of what we know about the way the ocean crust is constructed, about the history of ocean currents, and about climate change through the last 100 million years, comes from this extraordinary body of knowledge culled from the deep ocean floor. It has told us more—far more—about planetary function than any other single endeavour, and is still largely unknown to the public.

Going Back

If we want to reconstruct the oceans of the past and go back a modest span of time—say, to the heyday of the dinosaurs 100 million years ago—then, thanks to the endeavours of the Ocean Drilling Project, this is now almost straightforward. One can simply run the loop of time backwards, close up the segments that have recently opened, magnetic stripe by magnetic stripe and time-slice by time-slice. Running time backwards this way, the Atlantic Ocean progressively narrows and disappears: by some 200 million years ago, in the early Jurassic, it was entirely closed.

At that time, oceans that have now vanished, or are just remnants, become enormous. The Mediterranean Sea is expanded from its present shrunken state into the mighty Tethys Ocean of ancient times, as Africa pulls back from Asia and India tracks back south of the equator. The Alpine and Himalayan mountains, meanwhile, subside back into the original, uncrumpled shallow seas and coastal plains that they sprang from.

But to go yet farther back? After all, close up the present-day ocean and we are back just a couple of hundred million years, which is less than 5 per cent of the age of the Earth. How does one reach yet further back, to divine the shapes of yet more ancient oceans?

The detective work here is a little less straightforward, and therefore more fascinating. With virtually all of that ancient oceanic crust destroyed we need to make enquiries among their more durable neighbours, the continents, by tracking their movements. Once we know where the continents were, then the oceans must have been everywhere else.

Ancient continental positions can be tracked by measuring the orientation of preserved magnetic particles in the strata that still point towards where the North Pole used to be when those strata formed. This kind of information is patchy (the magnetic data can be overwritten if the rocks are heated too strongly, for instance), and it provides information only on the ancient latitude, not the longitude, of strata. Nevertheless, the magnetic information can be combined with other approaches, such as making a 'best fit' of the edges of ancient continents in what is essentially an Earth-sized jigsaw puzzle. These former continental edges themselves may show up as ancient, eroded mountain belts—crumple zones that mark where continents collided to destroy the ocean between them. Here and there, slivers of oceanic crust may have been rescued from subduction, caught up in the crumple zone and scraped on to the continental edge—where they may lie still, to be interrogated by geologists today.

For instance, one can follow the line of a 400 million-year-old belt of eroded mountains from Scandinavia, across Britain and Ireland, into Newfoundland (that used to be just next door to Britain before the Atlantic Ocean opened), then down the Appalachians. These, the Caledonian Mountains, represent the giant crustal scar tissue where there was a former ocean—the Iapetus Ocean—that is in many respects the forerunner of the Atlantic Ocean. Once several thousand kilometres across, it was swallowed up into the depths of the Earth: that is, the crust of the sea floor was swallowed up, while the water itself simply spilled across into other ocean basins.

One can occasionally still find slivers of this ocean floor—for instance at the small Ayrshire village of Ballantrae on the west coast of Scotland.[45] Here there are telltale signs of ancient ocean floor, some 480 million years old, which include basalt lavas with 'pillow' structures (the pillows formed by water chilling as the lava is extruded) and the fossilized ocean floor oozes that are associated with them. Ballantrae is a tiny sliver, a few kilometres long, of that ancient ocean that once separated Scotland from England and Wales. A little distance away, in southern Scotland, there are more oceanic slivers—but these just represent the ocean floor sediments (now hardened into mudstones) that were scraped off the basaltic ocean floor and stacked vertically—with uncanny gentleness, as though by a giant and expert plasterer—against what was then the edge of a Scottish/North American continent. This crustal scraping and plastering took place over a period of about 30 million years, as the Iapetus Ocean was disappearing. The underlying basaltic crust here disappeared completely: all of it was pushed down into the mantle (see Plate 2).

A little to the south there was another ocean, the track of which can be followed by another line of eroded mountains from southern Britain through Spain and Germany into south-west Poland. This ocean, the Rheic Ocean, essentially lay between Europe and Africa. It was a forerunner of the mighty Tethys Ocean, the demise of which gave birth to the Alps and the Himalayas, and of which the Mediterranean Sea is the shrunken remnant.

Oceans of the past, therefore, have gone through cycles of opening and closure, each cycle lasting a few hundred million years. This is often called the Wilson Cycle, after John Tuzo Wilson, a Canadian geophysicist and one of the great pioneers of the plate tectonics hypothesis. It is also commonly referred to via its mirror image, the supercontinent cycle; a term that reflects the innate terrestrial prejudice of human landlubber geologists. The continents here are simply passengers, being carried almost literally on the backs of the ever-moving ocean plates.

Today we can see oceans at all stages of their evolution. There are incipient oceans, those initial fractures that are not yet filled by the sea. The classic example here is the Great Rift Valley of east Africa: a crustal slab tens of kilometres across and a few thousand kilometres long, that has dropped by a kilometre or more along parallel vertical fractures. Magma is already rising along these fractures, feeding chains of volcanoes. Soon, geologically speaking, Africa will split apart and a new ocean will be born. Linking with the Great Rift Valley, and a little further advanced, there is the Red Sea. Here, ocean crust has been forming for some 50 million years and has widened to some 300 kilometres across.

There are also the fully grown oceans, including those that are still widening—like the Atlantic, where every year Europe moves some two centimetres farther away from America, and the Pacific, which is closing around the Ring of Fire. Then there are oceans that are nearly dead, with the Mediterranean as the prime example. An ocean that is so constricted can die more than once (as we shall see in the next chapter), but its *final* death will come a few tens of millions of years in the future, when a mountain range has finally taken its place. Once that happens, it will take hundreds of millions of years for the future Mediterranean Mountains to be worn down, so that the sea can sweep in once more.

We can now, more or less easily, wind the story back by something approaching a billion years, with reasonable confidence over where the continents and oceans used to be. Over that time there have been a couple of supercontinents on Earth, when most of the Earth's continental crustal masses had joined together—and hence when there was more or less unbroken ocean all around. Three hundred million years ago, there was the supercontinent Pangaea, surrounded by the mighty Panthalassa Ocean. The crustal fragments that came together, long ago, to make up Pangaea had, some half a billion years earlier, been combined in yet another supercontinent that geologists called Rodinia, surrounded by an ocean to which the term Mirovia has been given. Names, names, names—it is a very human trait to name objects, large and small, even when those objects are constantly changing, changing neighbours and partners, and moving on.

It can seem to be a never-ending process. It was first glimpsed by James Hutton, a landowner and savant in late eighteenth-century Scotland. Hutton was a fine product of the Scottish enlightenment, but he left Edinburgh in 1747, likely because he got a girl pregnant. This period of Hutton's life began with an indiscretion, then, but it did not dampen his enthusiasm for all things scientific and modern. In the time away from Scotland he completed his medical studies, honed his skills in agriculture, and developed a new process for welding metal. He returned to his family's farm at Slighhouses in the Scottish Borders in 1754. Hutton had also developed a passion for natural history, partly inspired by the Comte de Buffon's *Histoire Naturelle.* The wages of sin of his exile were a mind able to ponder the age-old secrets of the Scottish mountains, and to develop insights that would change how people viewed the Earth and its history.

Walking among the rock strata, Hutton saw that sea floors eventually became uplifted and crumpled to become mountain ranges—and that when these were eventually worn down, the sea crept back. In realizing the scale of time necessary for these kinds of things to take place, he stumbled upon the tangible reality of deep geological time. Indeed, he thought of the process as eternal and endless, considering the Earth to have 'no vestige of a beginning, and no prospect of an end'.

We now know, though, that there was a beginning to the Earth, some 4,567 million years ago, as its main bulk accreted from colliding planetesimals (astonishingly, this figure seems to be precise to the nearest million years). The question about beginnings therefore becomes more nuanced. Have the oceans and continents simply been engaged in this stately dance since, for instance, plate tectonics began, or has the nature of the dance (and indeed of the dance floor) changed?

More precisely: has the proportion of oceans and continents always been more or less the same? Or, have the continents grown and expanded on a world that used to be mainly ocean? Or have the continents shrunk and the oceans grown?

This has been—and remains—debatable. Current opinion is that, through Earth history, the continents have grown overall at the expense of ocean crust. One line of evidence is the general structure of the continents, which—to a very crude approximation—are made up of an ancient central core surrounded by younger mountain belts.

Think of Canada, say: its flat central expanse of the 'Canadian Shield' consists of the eroded roots of mountain belts that are 2–3 billion (and more) years old, while the youthful Rocky Mountains rise in the west.

The Rockies—simplifying greatly—may be thought of as representing sediment washed into ocean trench systems at the continental margin, sediment that was then scrunched by tectonic pressures and plastered on to the side of the continent. The oceanic slab, inching its way down to the mantle beneath the Rockies, also causes melting of deep-lying rocks: the ascending magma, whether erupted at the surface or cooling at depth as granite bodies, also adds to the bulk of the newly forming mountains. By such processes, continents, over time, will grow at their outer edges, to slowly reduce the proportion of the Earth's surface that is made up of oceanic crust.

However, it is not quite this simple. It is now becoming clear that plate subduction does not always create continental crust. It can also destroy it, via a process termed subduction erosion. Here, the descending oceanic plate acts as a giant rasping file, wearing away fragments of the overlying continent and dragging them down to destruction with it in the mantle. In any one place, the balance between subduction accretion (i.e. continental building processes) and subduction erosion varies depending on a number of factors, such as the roughness of the oceanic crust (i.e. how abrasive it is), the angle at which it descends, and so forth. Quite how important subduction erosion is remains unclear—and almost certainly its importance has changed over time.

There have also been attempts to chart the growth of continents in the deep geological past by such means as trying to work out the areas or volumes of crust of different ages, or by looking at the spread of ages of magmatic rocks associated with mountain building. This is easier said than done, because the ancient cores of continents are largely overlain by younger rocks, and because older mountain belts can be refashioned as new ones grow. Nevertheless, patterns have been discerned—although (of course) different studies have reached different conclusions. Some models of continental evolution seem to show more or less steady growth, while others suggest that continents grew (and ocean basins therefore shrank) in distinct pulses.

The Balance of Water

Given that there may have been, on that early Earth, more ocean crust and less continental crust, how might that affect the oceans that filled them? The evidence of the rocks hints that, even in those early days, the seas might have covered a good deal of the continents. That might be explained if the ocean floor was generally higher (relative to the continents) on a young Earth, to help the waters spill over on to the higher ground. However, if the 'precursor tectonics' operated more sluggishly, it has been suggested[46] that the ocean floors might have been lower and deeper than now. And if *that* was the case, to help flood the continents there may have been more water in the ocean basins—perhaps as much as twice today's amount.

We need to think, therefore, about where that ocean water might have disappeared *to*.

We have been speaking so far of the ocean basins: the receptacles, as it were, of Earthly water—as if the surface of our planet was nothing other than a succession of different types and shapes of bowls to fit that water into. That isn't quite the case. There is also a balance of water between that at the surface and that which lies very deep in the Earth, dissolved in the rocks of the mantle and making those rocks flow more easily in the slow subterranean currents that drive plate tectonics (we will leave for now the question of any continued transfer of water from the surface to outer space).

We have already pondered on how the Earth's water arriving from outer space might have been 'folded in' to the mantle. So, how might that process have continued, and what might be happening now? And what might have been the long-term effect of any transfer of water between the surface and the depths of the mantle?

When volcanoes erupt today, the main gas that is erupted is water vapour, in large amounts. In addition to the major dramatic eruptions, water seeps and bubbles more gently through the crust in hot springs. Then, there is the flux of water going back into the Earth. Some water re-enters the Earth attached to the descending plate in the subduction zone, in the upper part of the fractured,[47] waterlogged basaltic crust, as well as in the water-soaked sediment above. Not all of this water is dragged down into the mantle, because in subduction the downgoing slab is both heated and subjected to very high pressures (it is a little like putting some wet cloth through a heated wringer). Most of the water is simply squeezed back towards the surface, fighting its way up through the mass of downgoing rock and sediment. As it is forced back upwards through the rock, it can leave telltale marks of its passage in the form of distinctive high-pressure injections of water-rich mud that cut across the strata.

Eventually, the mineral-rich fluid bursts back through to the surface in the ocean trenches in the form of 'cold seeps', submarine mineral-rich springs that are a haven for communities of specialized animals adapted to such a source of nutrients.

Part of the water may be carried farther down by the slab, as far as the upper part of the mantle, before being released. These fluids then pass upwards through the mantle above the downgoing slab, reducing the melting temperature of that mantle material. This triggers the ascent of water- and silica-rich magma that can emerge as devastating explosive volcanic eruptions. Krakatoa, Mount St Helens, Mount Pinatubo, Aconcagua, Popocatépetl: all these volcanoes (and many more) are of this type, fed by the magma rising from the 'wedge' of mantle that lies above downgoing crustal plates.

Yet more fluid travels farther down with the slab into the mantle, to be released, many millions of years later, as the subducted material is remelted to produce basaltic magmas at mid-ocean ridges. Hence, there are complex but substantial pathways of water between the surface and the depths of the Earth.

There is a large question here. After the initial influx and (presumed) 'folding in' of water, what has been the overall trend of water between the Earth's surface and its interior? Has water, on balance, been taken out of the mantle, causing the oceans to grow in volume? Or has water been continuously folded into the mantle, causing the surface ocean waters to further diminish? Simply by considering what we know of the inputs and outputs of water (the error bars are large), there seems to be more going into the Earth's interior from the oceans than is coming out. If so, that has consequences for the far future (see Chapter 8). From the geological evidence, though—that is, from clues in rocks and strata—it is hard to tell.

Using the proportions of land to sea (which we explored earlier in this chapter) is a crude measure, because there is the changing depth of oceans to consider too. We simply have to use the evidence of the continents as best we can, exploiting these permanent, buoyant structures as a kind of dipstick to see how shallowly or deeply they have been submerged in the past. In effect, one has to track past shorelines as they advanced inland on to higher ground, or receded, leaving former shorelines high and dry.

That is easier said than done. One problem is that sea level has bobbed up and down in the past for all kinds of reasons—as land ice has grown and melted, say. Twenty thousand years ago, when the Ice Age was at its height, sea level was some 120 metres lower than today (while melting all the ice that we have today would raise sea level by some 70 metres).

On longer timescales, mid-ocean ridges have grown larger or smaller as sea floor spreading has accelerated or decelerated globally in response to variations in the deep currents within the mantle. Larger ridges displace more water from the ocean basins to spill over the continents, while smaller ridges allow water to settle back into ocean basins that have become deeper and larger.

One has to try to resolve all of these short-term trends, and also to try to distinguish them from those apparent sea level changes caused by the land being tectonically raised or lowered. In any individual succession of rock strata, the patterns caused by purely local changes of sea level caused by tectonic uplift or downwarp of a landmass can look exactly the same as those caused by the global sea level rising and falling.

To disentangle all of this, one needs to correlate the sea level signals in strata around the world. This can help tell which sea level changes take place simultaneously and so are global, and which take place in one region but not in others and are due to sections of crust rising or falling. To work out this kind of history a precise time framework is needed, so that one does not match up a 100 million-year-old sea level rise recorded in one place with a 110 million-year-old sea level rise in another. With that kind of miscorrelation the end result is, geologically, gibberish.

Geologists are constantly aware of the gibberish factor, and of the fact that geology as a science is blessed—or cursed—with a terrible abundance of time. Therefore, they go to great lengths to pin down time in rocks as precisely as possible, using any kind of evidence that they can lay their hands on. This can range from the natural radioactivity of volcanic crystals to subtle changes in the patterns of isotopes of various chemical elements. The oldest method, though, is still in many ways the best and the most widely used in practice—the use of fossil animals and plants, exploiting the evolution and extinction of different fossil species as unique time-markers in rock strata. This works magnificently, but only for a little over the last half-billion years since large, complex, hard-shelled or bony multicellular animals appeared on Earth (see Chapter 6)—and that is not much more than 10 per cent of Earth's history.

Then, as our uncertain starting point, there are those hints of early, more capacious oceans, and of frequently flooded continents too, some 3 billion years ago—given that there seems to be an abundance of basalt that has erupted underwater in continental settings (such basalt develops particular structures: the 'pillow' shapes we noted at Ballantrae).

It may be, therefore, that the oceans have diminished, perhaps halved in volume, since their beginning. Size is not everything though—quality matters too. We have other aspects of the oceans to consider, such as: what makes them salty?

4

The Salt of the Earth

Before oil, there was salt. It bankrolled empires, determined the fate of kings—and kept people alive through the winter. Join the throngs of tourists and go to Wieliczka in southern Poland, for instance (Fig. 6). You descend along rough-hewn wooden ladders into an underground domain where kilometres of tunnels have been dug through salt-bearing strata. There are crumpled layers of white salt, grey salt, brown salt, and red salt. There are salt crystals and salt stalactites, and shrines and chapels too—carved out of the salt by human hands, with salt altars and salt saints and salt Madonnas.

For over half a millennium, money from the salt mines funded military campaigns of what was then the largest nation in Europe. It built cathedrals, and kept poets from starving. It changed the landscape for the winning of the salt, clearing the region of forests to burn the fires that evaporated the brines. It sparked in human hearts the familiar combination of great generosity and single-minded rapacity—greed for the white gold broke empires as well as making them. In 1666 the ambitious nobleman Jerzy Lubomirski fought the king for the rights to the salt and won. The loss of power from the centre is seen as a decisive turning point in the decline of a nation. In successive wars, the divided country shrank back. By 1795, Poland had disappeared from the map of Europe, swallowed by its neighbours.

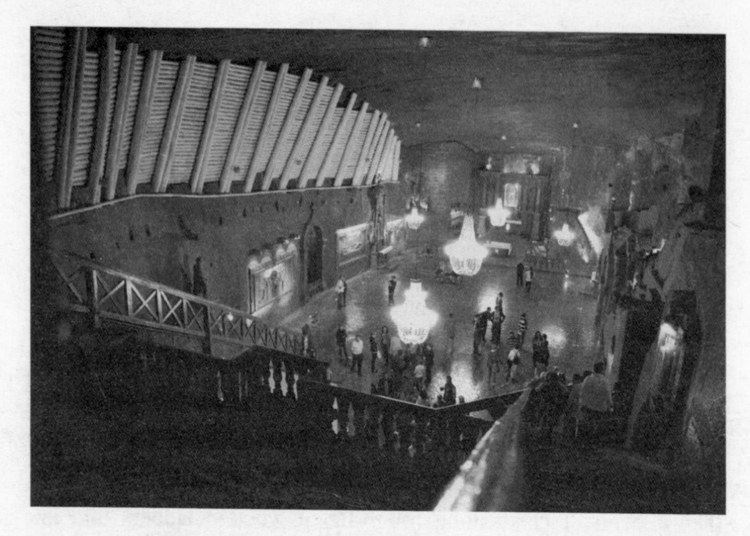

FIG. 6. Fossilized Miocene seas in the Wieliczka Salt Mine, Poland. St Kinga's Chapel is hewn from the fossil salt, and is 31 metres long and 15 metres wide.

For everyone needed salt. In the days before refrigerators and plastic packaging and tin cans, salt was needed to preserve meat, fish, butter, cabbage. It was used in tanning leather, in making glass, in making gunpowder. Rich or poor, everyone needed it. The men who controlled its supply were the Rockefellers of their day. They wielded true power, and jealously guarded their privileges. For salt was not easy to obtain. Winning salt from the sea by evaporation, in salterns, was long practiced, but slow and cumbersome. An underground supply of millions of tonnes of salt like Wieliczka, therefore, was better than a goldmine.

Neither the generations of miners, nor the merchant barons, knew that the crazily twisted layers of rock salt that they excavated marked the long, slow dying of a sea, many millions of years ago. That demise, looking further, is part of an even-longer process that governs the saltiness of the oceans and determines the chemical environment of all the organisms that live in them.

The Salt in the Sea

Seawater is salty. What, then, is the saltiness? If we take the water out of a litre (or a kilo) of ordinary seawater, then what we have left is some 35 grams—an ounce and a half—of solid matter.

Taking this solid stuff into its constituent particles (that is, its ions), most of these—more than 85 per cent of this residue—are of chloride and sodium, and when crystallized together these make up sodium chloride, or common salt. Then there are just six others that make up most of the rest: sulphate at a little under 8 per cent, and magnesium at just under 4 per cent; there is also calcium and potassium at a little over 1 per cent each, and bicarbonate at a little under half a per cent, and bromide at a fifth of a per cent.

That's reached over 99 per cent. Everything else is crammed into the remaining fraction of 1 per cent. There is a lot of that everything else. There is silica and iron and lithium and boron and tin and tantalum and cadmium and mercury and arsenic and . . . there are measurable amounts, indeed, of nearly every element in the periodic table. Water, as we have said, is a marvellous solvent.

How much, though, of the real rarities? One might consider gold for instance—that most perennially fashionable of elements. There are, appropriately enough, some fine cautionary tales to be told of gold in seawater. One might start with Svante Arrhenius, a genius at chemistry—and indeed a recognized genius, winning a Nobel Prize in 1903. We will meet Arrhenius again in Chapter 9, as he explores other worlds. On Earth, though, he considered (among many other questions) the amount of gold in seawater. Around the beginning of the twentieth century, he carried out analyses that gave a figure of

6 milligrams in every tonne of seawater: not a large amount, but not trivial either if a means could be found of extracting it.

Enter, two decades later, Fritz Haber—another giant of chemistry, and indeed another Nobel Prize winner. Haber is a man who, literally, allowed humanity to grow. He invented an effective means to turn nitrogen from the air—a very unreactive gas—into ammonia, the feedstock for nitrogen-based fertilizers. About half the protein in your body (and in ours, too) is built around nitrogen fixed by the Haber process. Without it, half the world would starve. With it, there will naturally be consequences, which we will explore. But for now we can simply turn to one of the more curious episodes in the life of Haber, a man whose life combined high achievement and deep tragedy.[48] In 1920 he was a famous German scientist in a Germany crushed under the weight of Allied demands following the end of the First World War: a total of 132 billion gold marks had been demanded as war reparations.

Haber wanted to save his country from penury, and remembered Arrhenius's work on gold. He repeated Arrhenius's analyses and came up with similar figures. A golden future beckoned. With colleagues, sworn to secrecy, he set sail on the oceans to follow up the work, collecting more samples and working on ways to extract the riches from the oceans.

The gold stubbornly refused to appear. What was wrong? They carried out more analyses, more carefully—the amount of gold went down and down with successive tests. Their dreams of wealth, and of rescuing their country, fell apart. Instead of 6 milligrams in a tonne of seawater, there turned out to be about *a hundredth* of a milligram—nowhere near enough to extract for profit, not even for a chemist of Haber's powers. What had gone wrong?

The answer was laughably simple: contamination. Gold in the wedding rings that the scientists were wearing, in their spectacle frames, in their gold teeth; trace amounts of the metal were even present in the glassware and the reagents they were using. When one is trying to measure minuscule amounts of something, then even tiny amounts of contamination can ratchet up the errors by orders of magnitude. Slipshod practice by scientists who should have known better? These were early days, we must remember. Rather, it was progress in analytical precision, painfully attained.

Of other substances in seawater, some are a little surprising, taking our 30 kilograms or so of sodium chloride and one hundredth of a milligram of gold, per tonne, as measured on a scale of abundance. Take iron for instance. It is abundant in the Earth's crust and a major component of many of the common rock-forming minerals. In seawater there is a mere 3 milligrams per tonne—something that has had profound effects, and seen profound changes, in time and space within the Earth's oceans, as we will see later in this chapter and in Chapter 6. There is more molybdenum (about 1 gram per tonne) and rubidium (more than 10 grams per tonne) dissolved in the waters of the sea—although less aluminium (about 1 milligram per tonne).

And so we can go on, running through the elements—even those present in smaller amounts than gold. There are quite a few here: rhenium, osmium, ytterbium, thorium, and more. The saltiness of the sea, indeed, has almost infinite variety. What is a little surprising, perhaps even disconcerting, is that the chemistry of the sea now bears little resemblance to the chemistry that is being washed in by rivers. There is a mismatch here, and that does demand explanation.

The Salt Supply

Water, one may repeat, is a marvellous solvent. As it travels its powers of dissolution may, indeed, increase. Drifting in the atmosphere as fine cloud droplets and falling through the air as raindrops, it absorbs carbon dioxide, sulphur dioxide, nitrogen dioxide, and other gases that dissolve within it. That rainwater is, therefore, a cocktail of dilute acids. 'Pure' rainwater is mildly acidic, with a pH between 5 and 6 (7 is neutral). Rainwater near thunderstorms or erupting volcanoes (or, these days, industrial cities) can have a pH of 3 or below, to become as sour as lemon juice.

As it falls on to the ground it permeates the soil. There, the dead and decaying plant and animal matter in humus adds its own complement of organic acids to the mix. Fully armed, the water can trickle on and set to work. It begins to attack the mineral structures of the rocks that it seeps past. Most of these minerals were forged in the depths of the Earth, at temperatures of several hundred degrees Celsius or more and at high pressures. At the cold, damp surface of the Earth, these minerals are distinctly uncomfortable—or, more technically, metastable.

Keep these high-temperature, high-pressure minerals dry, though, and they can last forever, no matter that they are so far out of their comfort zone. The beautifully preserved igneous minerals in the rocks brought back from the bone-dry Moon in those long-gone days of danger and high adventure bear testament to this. Billions of years old, they still retain their pristine structure, as fresh and unaltered as the day that they crystallized.

However, on a moist, gently acid surface such as that of the Earth, such minerals are quickly dismantled. The regular atomic frameworks of the silicate minerals—olivines, pyroxenes, feldspars, micas—are broken down. The solid debris that is left is a variety of clays, the main ingredient of that wonderful substance, mud. What concerns us, though, is the chemistry that goes into the solution and travels with the water on its journey through soil, into streams, into rivers, and then into the sea.

The chemistry of river water varies from place to place. It depends on what kind of minerals, of what chemical composition, the water flows past. If one were to mix river waters from different parts of the world together into some sort of broad average, though, then one would see a general pattern. For a start, there are almost no rivers that are anywhere near as salty as the sea. The salt contents are typically less than 1 per cent of that of seawater. Freshwater may not be pure, but it is nowhere near a brine. That is understandable. That river water derives, more or less directly, from water that has evaporated from the sea—and when even a concentrated brine evaporates, virtually all that goes up into the air is simply... water—in effect distilled water. That is the starting point, to which the acids and dissolved minerals are added.

So—is river water simply dilute seawater? It is not close to being even that. Sodium and chloride make up some 85 per cent of the salts in seawater, as we have seen. In river water, together, they amount to less than 15 per cent. Bicarbonate makes up half a per cent of seawater salt—but nearly one-third of the dissolved solids in river water. There's 1 per cent or so of calcium in seawater, but over 15 per cent in river water. And so on. The chemistry of seawater and river waters are very unlike each other—and yet the rivers are the main source of salts to the sea.

This puzzling circle can be squared—and relatively easily, too. One just needs to consider what happens to those salts when they reach the sea. For they can have many different kinds of journey within the ocean waters. Each type of dissolved substance has its own trajectory, and it is that which gives us our familiar seawater. Chemistry is one control here, while life is another. We can turn to these first, because they have come to work steadily and systematically, and it is nice to see the world work in such an ordered fashion (later we will talk of how saltiness also relies upon tectonic accidents).

The Removal

It is what goes out, quite as much as what goes in, which controls the saltiness of the sea. For some elements simply do not dissolve easily, even in that universal solvent, water. Take iron for instance, in today's oceans (in yesterday's oceans things here were quite different, as we explore in Chapter 6). Wherever there is oxygen—which is nearly everywhere in ocean and river waters—the iron converts to its trivalent form, Fe^{3+}. This is almost insoluble, and simply converts into fine particles of hydrated iron oxide, or rust, which falls to the floor of the river or (if the iron has made it that far) the sea. There are many elements that are very poorly soluble in water—cobalt, nickel, tin, mercury, and gold for instance—and so their contents are very low, and their contents in seawater may be measured in those milligrams, or fractions of a milligram, per tonne.

Interestingly, the amounts of most of these *trace* elements in seawater is even lower than their simple solubilities, as measured in a laboratory, would indicate. Hence, they are not simply being removed from the water by precipitation. There are other processes at work. Some can be purely inorganic—these elements can be adsorbed, or 'stick to', the surfaces of clay or rust particles where the water is muddy, for instance, and be carried with them to the sea floor. In others biology is at work; planktonic organisms can absorb these elements (many are micro-nutrients) out of the seawater. They can then be passed from prey organism to predator organism up the food chain, until something somewhere along the chain dies and falls to the sea floor, again taking those elements with it. Or the elements may leave the water column yet earlier, in sinking faecal pellets or as discarded gelatinous feeding webs that travel as 'marine snow' (also known as the 'faecal express') to the sea bottom (see Chapter 6).

Life is thus a great regulator of the chemistry of the oceans—and not only as regards trace elements. Living tissue is built of carbon, nitrogen, and phosphorus, and if it has bones or teeth or scales then yet more phosphorus, together with calcium, is needed to build that skeleton. All of this derives, directly or indirectly, from seawater.

Then there is the skeleton-former of choice for molluscs, corals, stony algae, sea butterflies, and those giant armoured protozoans the foraminifera that live both at the sea surface and on the sea floor. This is calcium carbonate, and its wholesale extraction from the sea floor by these organisms forms, and buries, enormous masses of limestone, thus taking its ingredients—calcium and carbon—out of the water and converting them into rock. The chemistry may be simple enough, but the architecture of these calcium carbonate shells and skeletons is dazzlingly sophisticated. The elegant shape of a scallop, cowrie shell, or coral that one can appreciate with the naked eye is beautifully engineered in itself. But take a scanning electron microscope to a tiny portion of such a shell, or to the whole skeleton of a foraminifer no bigger than a sand grain, and magnify it a few thousand times. Then there appear marvellously shaped and interlaced crystals of the shell mineral encased (in life) in an organic matrix. More finely sculpted and assembled than a Fabergé egg, each is a reminder of the almost storybook complexity encapsulated within, seemingly, the most common and humble of biological constructions.

This particular import–export pathway is modulated by life—but is not dependent on it. So much calcium and carbonate is carried by rivers into the sea that the bulk of the ocean waters are saturated in calcium carbonate, and it can precipitate out of the seawater of its own accord. On a hot sunny day off the Bahamas, for instance, especially if a drying wind is blowing, the surface of the sea can suddenly turn milky white. 'Whitings', as these are called locally, are made of millions of tiny needle-like crystals of calcium carbonate that have spontaneously precipitated out of the seawater—they then drift slowly to the sea floor to form a layer of white carbonate mud. If our planet was suddenly turned lifeless, whitings would be forming all the time as part of the natural regulation of the calcium carbonate component of the saltiness of the oceans.

Chemical Lifetimes

There are those substances that, unlike iron or gold, can be dissolved in enormous amounts within seawater. We have the chemical giants of the ocean, chloride and sodium, and, some way behind, potassium, magnesium, and sulphate. Water can absorb immense amounts of these substances, much more than the modest amounts of carbonate and the tiny amounts of iron and gold. Far more, in fact than there are in today's seas: think of the landlocked Dead Sea, so dense with dissolved salts that it is impossible to sink, but difficult to swim. The question of why the oceans are not more like the Dead Sea is something we will come to shortly.

The compounds one can make of these ions, such as sodium and potassium chloride, are so soluble that no organism can find a means of extracting these from the seawater to form a skeleton, despite the abundance of such potential raw materials in water. Bones of common salt would simply dissolve away, and no sophisticated cellular biochemistry has been evolved that might prevent this.

These ions therefore, once in the water, simply stay there—for a very long time. There is a concept that can be used here that is as vivid as it is useful. It is called the *residence time,* and it simply means the average time a particle of anything stays within a particular system. Thus, the residence time of an ion of calcium say, in the ocean, once washed in by rivers, is about a million years. That is quite a long time—by human standards certainly—but sooner or later the calcium will combine with another ion, often of carbonate (perhaps within a coral polyp or the tissues of a mollusc), to form part of a calcium carbonate crystal, and thus be taken out of the seawater system. As such, the average calcium ion in the oceans has been there since before the human species evolved, and since the times when mammoths and woolly rhinos roamed the Earth.

For the common seawater ions, residence times are of truly geological durations. For potassium it is about 7 million years; for magnesium, 10 million years; for sodium, 70 million years—and for chloride the residence time is a simply gigantic 100 million years. The average chloride ion, therefore, will have been carried around the world again and again on the ocean currents since the heyday of the dinosaurs. These are average figures, we must recall. Some of these chloride ions were washed into the sea last week. Others will have been in the sea much longer. A single drop of seawater will contain many chloride ions that have drifted there since the trilobites roamed the sea floors, half a billion years ago.

These enormously long times mean that these common ions have been well and truly mixed and homogenized in the oceans, the 'mixing time' of which (the average time it takes for water particles to circulate around the oceans) is about 1,000 years. That is not to say that their concentrations are identical everywhere (that character will become very important in the next chapter, where we discuss how ocean currents work), but the relative proportions of these ions are the same, as are the proportions of any stable chemical isotopes of them.

For the trace elements in seawater residence times are typically much shorter, mainly because they quickly drop out of solution, being crystallized biologically or chemically or by a combination of the two. For iron, the residence time is a mere 200 years—so most of the tiny amounts of dissolved iron in the oceans date back only to the Industrial Revolution. The residence time is not much greater for such elements as aluminium, lead, and zinc—and also less common (but useful) ones such as the rare earth elements, including cerium, lanthanum, and neodymium.

These very short residence times can be useful for geologists. Being much shorter than the mixing time of the oceans, such elements can act as oceanographic tracers, illuminating such processes as the present (and past) inputs of water from different rivers, or the changing patterns of ocean currents. To use this feature in practice, one needs an element that can provide a distinct 'fingerprint', because simply measuring the abundance of an element in seawater or in sediment does not provide precise enough information. Such an element is neodymium, which has different isotopes, the proportions of which change from place to place. The careful measurement of neodymium isotopes in ancient seawater sediments has told stories of how the outflow of water from the Nile into the Mediterranean has changed over the climate vicissitudes of the Ice Ages,[49] or how the Gulf Stream has changed its behaviour over the past few million years.[50]

So seawater is a complex cocktail of chemical elements, some of which can enter and exit the system almost on the scale of a human lifetime, while others travel in cycles about as long as those of the drifting continents. Given the enormous capacity of the ocean to absorb ions of sodium, potassium, and chloride, it could potentially become a brine denser than the Dead Sea, given that rivers have been washing these salts into the sea for more than 4 billion years.

So why are the oceans not entirely saturated with salt? One answer is to be found among a string of accidents. Wieliczka was one, a very long time ago. The most dramatic example, though, is centred in Messina, Sicily, by the beautiful blue Mediterranean Sea—or perhaps that should be by the *presently* beautiful blue Mediterranean Sea.

Death of the Mediterranean

Messina has history. It has been invaded successively by the Carthaginians, the Syracuse army, the Mamertines, the Romans, the Goths, armies of the Byzantine and Arabian empires, the Normans, the forces of Richard the Lionheart, the Spanish army (among them Miguel de Cervantes, author of *Don Quixote*, who recovered in the local hospital from wounds sustained in battle), and by Garibaldi's army. It has been devastated by the plague (it was the gateway for the Black Death into Europe) and by earthquakes.

Another catastrophe that happened far longer ago was discerned by geologist Giuliano Ruggieri in the 1960s, as he walked among the strata exposed in the crags and cliffs of the rugged local landscape. Among them were the picturesque cliffs and bay of Eraclea Minoa named—local legends differ—possibly after Heracles (Hercules), or alternatively after Minos, the ancient king of Crete. In those cliffs there are beautifully displayed strata from the late Miocene and Pliocene epochs. Some 5–8 million years old, they represent a deep-sea floor of that time, now pushed up above sea level by the tectonic forces that are squeezing the Mediterranean shut. Among them are layers of gypsum—calcium sulphate—one of the minerals that forms when seawater evaporates.

Perhaps, said Ruggieri, the whole Mediterranean then dried up. It was a prophetic statement. It looked back, too, to earlier legends. Pliny the Elder had recounted a story that there used to be mountains blocking the Straits of Gibraltar, until Hercules (that man again!) dug a passage so that the Atlantic waters could flow in. In the 1920s H. G. Wells thought on how nature, rather than assiduous super-heroes, might have acted to similar effect. He noted that the global sea level dropped by over 100 metres at the height of the last Ice Age. That would block inflow at the narrow Straits, he thought, and make the Mediterranean largely dry up—before the post-glacial sea level rise brought the sea back in. His logic was good but his bathymetry was out, as the channel stayed deep enough throughout the Ice Ages to maintain constant connection with the ocean. Robert E. Howard, meanwhile, just had fun with the idea, having his own superheroes Conan the Barbarian and Red Sonja carry out their feats of derring-do in the far-off Hyborian Age, across a wide landscape centred on a dry Mediterranean.

The true story—and the huge scale—of the Mediterranean desiccation puts all of these fictional and legendary accounts in the shade.[51] It was yet another of the succession of ocean wonders revealed by the carefully manipulated drill-strings of the *Glomar Challenger*, as it sailed in 1970 to discover what lay beneath the floor of the Mediterranean Sea. On board were three scientists, Kenneth Hsü, Bill Ryan, and Maria Bianca Cita—all of them on the track of the oldest stories of this most history-packed of seas. What they found, as core after borehole core was pulled out, was salt—prodigious thicknesses of salt. The salt layers beneath the Mediterranean—once the borehole evidence was integrated with seismic images of the strata—were shown to be up to 3 *kilometres* thick. This was not some minor drying event. This was wholesale destruction of an enormous inland sea, with repercussions that were literally worldwide.

Meticulous analysis of the strata over the succeeding years has revealed the sequence of events. A little under 6 million years ago, connection of the Mediterranean with the Atlantic Ocean was broken off, as tectonic changes led to a barrier forming at the Straits of Gibraltar. Given the dry climate, and the small river inflow, it may have taken only a few millennia or so to reduce the beautiful blue Mediterranean Sea to a blindingly white salt desert—*quite* unlike Conan the Barbarian's sylvan Hyborian landscapes—with scattered lakes of concentrated brine. The bakingly hot, devastated landscape lay up to 5 kilometres below the level of the global ocean (which had been raised by about 10 metres as a result—because all that water had to go *somewhere*). The rivers flowing in, adjusting to the new geography, carved deep Grand Canyon-like gorges into the freshly exposed landscape.

There is something to explain here, which is that even total desiccation of the Mediterranean Sea would only give rise to salt a few tens of metres thick, and not up to three kilometres. What clearly must have happened is that Atlantic waters periodically flooded in to replenish the sea—only to dry out in turn, leading to yet another layer of salt building up on the surface. This has been explained in terms of competition between the tectonics that built the barrier at the Straits of Gibraltar, and the forces of erosion that wore it down.[52] And so, over the next 700,000 years, there were successive brief influxes of Atlantic water. Each influx soon dried out, and so the salt built up and up.

In total, an estimated 1 million cubic kilometres of salt lie buried beneath the present-day Mediterranean Sea floor.[53] That is more than 50 Mediterranean Sea's worth—and was about 5 per cent of the salt in the global ocean. Thus, within the space of less than a million years, the ocean salt content dropped by that 5 per cent forever—or at least for as long as the salt stays locked up in those strata beneath the sea floor.

What would have been the effect of this salt transfer? Clearly, within the area of the Mediterranean virtually all of the marine biology would have been killed off as the area became a toxic desert. But outside it? The salinity reduction would have had wider effects, in raising the freezing point of water and making it easier to form sea ice, for instance. Did this have any effect on global climate? No one yet knows.

The rebirth of the Mediterranean was even more sudden than its death. Above the thick salt deposits there is a sudden change to normal marine sediments, this happened simultaneously across the basin 5.3 million years ago. The Gibraltar barrier, then, must have finally broken completely to allow the Atlantic water to rush back in, in what geologists term the Zanclean flood. Estimates of the refilling, based upon deep scours still preserved around the Straits, suggest that it took only some two years to accomplish.[54] If so, it must have been one of the most spectacular floods in Earth's history. Perhaps the thundering waters were witnessed, in mute terror, by the ancestors of today's Gibraltar apes. Let's hope they had a safe perch, high on the island's peak.

Saltiness through Time

As we go back through Earth's history, other salt-depositing events are encountered, each one an accident of geography, tectonics, and climate. The Wieliczka deposits are about 14 million years old and represent a smaller event—the blocking off of one arm of the Mediterranean. As the Atlantic Ocean was first forming, its narrowness led to intermittent isolation and drying—and more salt deposits. In the Permian Period, over a quarter of a million years ago, a desiccating sea called the Zechstein Sea covered much of Europe, from eastern England into Germany—and another one stretched across Texas and New Mexico. Each of these events drew enormous amounts of salt out of the sea as it formed, and each significantly reduced the ocean's salt content.

The salt layers now lie at depth, and they make strange strata. Rock salt for instance, when put under pressure, slowly flows—like an enormous underground salt glacier. It flows upwards, because the salt is less dense than silicate rock, into pipe-like structures termed diapirs that can be hundreds of metres across and many kilometres high. They punch through the overlying strata, fracturing and dislocating the rock layers. Oil geologists love such salt diapirs, because oil often becomes trapped at the contact between layers of permeable rock and the impermeable salt.

The sea's overall saltiness, then, is governed by a balance between import by rivers into the sea and export of salt minerals in rock strata. The former is a more or less steady, incremental process globally; the latter is fitful, and dependent upon those tectonic accidents. Which process, over geological timescales, is winning?

It used to be thought that the world's oceans have been getting slowly saltier through geological time. Indeed, the saltiness of the oceans was once used as one of many ingenious attempts, before the discovery of radioactivity, to estimate the age of the Earth. This was done by simply estimating the rate of salt inflow from rivers and assuming that the greater part of that salt simply stayed in the sea (this was before the enormous scale of the 'giant salt' deposits such as those of the Mediterranean was appreciated).

However, to create giant salt deposits needs, above all, landscape—or, more precisely, a good balance between land, shallow sea, and deep sea, to give maximum opportunity for the kind of tectonic accidents that can dry out a good-sized sea. In the depths of Precambrian time, in the Hadean and early Archaean eons, there seems to have been less dry land and more ocean, partly because there may have been more water (before part of the ocean got dragged into the mantle along with subducting tectonic plates) and partly because continents were fewer and smaller.

If that was the case, then salt import into the oceans may have dominated over salt export in the early part of the Earth's history, to allow the oceans to become saltier than today—perhaps up to twice as salty in some estimates. Then, when a different continent–ocean balance was established, salt could begin to be buried in large enough amounts to begin to reduce the Earth's salinity to current levels. There is a hint of this kind of scenario, in that giant salt deposits seem to be rare until relatively late in Precambrian times.[55] Few rock strata survive from those early days, though, so the evidence is slender.

Even without an ocean drying up completely, Messinian-style, reduced flow from one part of the world ocean to another might lead to large differences in salinity from one part of the world to another. Offshore from eastern North America beneath Chesapeake Bay, there is a buried crater some 50 kilometres across from a meteorite that, 35 million years ago, ploughed into strata of Early Cretaceous age (about 100 million years old). Boreholes put down into the crater showed that the strata within it are bathed in very strong brine, about twice as strong as modern seawater. Perhaps this was from the underground dissolution of buried salt deposits? None, though, are known nearby. Or by heating and evaporation related to the giant impact? That, physically, could not work, and the brines extended well beyond the crater. The only reasonable interpretation was that the buried brines represent Early Cretaceous seawater of the North Atlantic—then much narrower than today and sufficiently hemmed in by the Americas, Europe, Asia, and Africa to develop its own, concentrated version of seawater.[56]

There must have been other kinds of changes too. It is not only the quantity of salt in the oceans that matters. The quality of that salt is important too. In fact, it controls life and death.

Limestone Patterns

Well-mixed as the seas are today, ocean chemistry is not quite uniform. It varies across the wide surface of the oceans—from the coastlines, where the water is full of trace elements of all kinds that help life grow in abundance, to the ecological deserts in the centres of the oceans.

Ocean chemistry also varies from top to bottom. The surface waters of the ocean are relatively alkaline because carbon dioxide (that becomes carbonic acid when dissolved in water) is used up as myriad tiny planktonic algae photosynthesize and grow. In those sunlit waters it is easy for some of those planktonic organisms to extract the calcium and carbonate ions[57] with which the water is saturated, and make their intricate, jewel-like calcium carbonate skeletons. As these organisms die, their skeletons fall to the sea floor in their billions to pile up as carbonate oozes that are limestone strata of the future. As they fall, they take with them the iron, nitrogen, and phosphorus in their dead tissues. The sunlit surface waters are thus depleted in these elements, which puts a serious brake on further biological growth.

However, to build future limestone strata those tiny skeletons must not sink too deep, for if they do they will enter a new realm and simply disappear. In the cold, dark depths of the oceans, about 4 kilometres below the sea's surface, the water is both cold and old. Hundreds of years have elapsed since the water has been at the surface—it is charged with carbon dioxide from decaying animal and plant tissues, and part of this gas dissolves to form carbonic acid. As they enter these deep, more acid waters the calcium carbonate shells simply dissolve and disappear, even before they hit the ocean floor.

There is therefore a kind of snow line known as the carbonate compensation depth (or CCD) in the oceans, above which pale oozes made of microscopic calcium carbonate skeletons accumulate. Below this only fine, insoluble sediment can very slowly build up, which is made of material such as windblown desert dust, tiny particles of volcanic ash, and the tiny silica skeletons of such organisms as radiolaria or diatoms. One of the fundamental boundaries in the oceans, the CCD varies from place to place. It is generally shallower in the Pacific than the Atlantic, because Pacific deep waters are more sluggish and therefore older and more acid. And above productive areas of surface water the carbonate snow line can be pushed downwards, because the rain of skeletons from above can overpower the ability of the deep water to dissolve them.

The CCD has also changed through geological time. There have been, for instance, times in the past when the atmosphere was suddenly invaded by large amounts of carbon dioxide ('suddenly' here means over tens of thousands of years, by comparison with what is happening today: see Chapter 7). One such event took place 55 million years ago, perhaps as a result of an intense burst of magmatism associated with the opening of a part of the Atlantic Ocean. As well as warming the world it acidified the ocean, so that the sinking carbonate skeletons dissolved sooner and the CCD rose by about a kilometre.[58]

There were other, more subtle changes in the oceans that affected the way in which calcium carbonate precipitated out of the seawater. One involves changes in the relative proportions of calcium and magnesium in the ocean waters, where the balance between the two has swung to and fro over at least the past half-billion years—and probably since the oceans first formed. This ratio has proved surprisingly difficult to measure in ancient rocks, not least because the chemical patterns are prone to change as the strata lie buried deep underground over millions of years. However, traces of these ancient chemical patterns may be preserved in particular types of calcium carbonate (such as that in the crystals of sea urchin skeletons), and also in what seem to be true samples of ancient (albeit concentrated) seawater, in fluid bubbles trapped within fossil crystals of rock salt that once crystallized on a sea floor.

These patterns show oscillations between times of magnesium dominance (such as now) and calcium dominance (such as 100 million years ago in the Cretaceous Period, when dinosaurs roamed the Earth). This magnesium–calcium oscillation seems to reflect different plate tectonic modes of the Earth. When sea floor spreading is vigorous, the abundant, newly erupted underwater basalts will absorb magnesium from the seawater leaving it rich in calcium. When sea floor spreading slows the magnesium absorption slows, leaving the seawater richer in this element.[59] This slow chemical oscillation seems to exert a strong influence on global climate,[60] and so on the course of life's evolution, both in the oceans and on land.[61] Life, to reciprocate, has also changed the pattern of ocean chemistry, as we shall see in Chapter 6.

There has been another sea change linked with life's evolution on Earth that was just as profound: the slow spread of oxygen through the early oceans altered them fundamentally. Over a billion years and more, the early Earth possessed bizarre iron seas and sulphide seas. When those were swept away, our planet—as *almost* every animal and plant would agree—became a better place.

Iron Seas, Sulphide Seas

There are different kinds of saltiness. The world's oceans, today, have a chemistry dominated by sodium, potassium, chloride, and sulphate. However, if you travel in desert regions and stumble across dried-up lakes there you may likely find their saltiness to be quite different, as in the bitter lakes where sodium sulphate is dominant because the rivers that feed them bring in specific, local chemistries.

You don't usually have to travel so far to see a different kind of saltiness. Take a walk in coal-mining districts, or in an area with thick peat bogs, and you will likely see springs here and there bringing water that, when it reaches the surface, becomes orange, opaque, and slimy: rather unpleasant, in short. This water is iron-rich, and you are looking at a faint echo of the Earth's earliest oceans.

Water that contains no oxygen—water that was typical of the first 2 billion years of the Earth's oceans—can contain large amounts of dissolved iron, many times more than can be found in both typical freshwater and seawater in today's oxygen-charged world. When oxygen comes into contact with such water, the iron immediately forms an insoluble hydroxide—a particle of rust—and becomes a part of that orange slimy stuff you can see in the iron-polluted streams.

Early oceans contained iron in abundance, and on the sea floors from early Archaean times, more than 3 billion years ago, thick layers of iron oxides accumulated that today form giant iron ore deposits which provide nearly all of the iron and steel that we use today. It used to be thought that this was a reflection of early oxygen-producing photosynthesis by marine cyanobacteria—but iron oxide deposition started in the sea over a billion years before the Great Oxidation Event saw oxygen enter the atmosphere, to rust and redden terrestrial land surfaces. The formation of these enormous, ancient iron deposits is therefore now thought to relate to a primitive form of photosynthesis that did not produce oxygen. Nevertheless, it was an effective means of slowly beginning to take out some of the enormous amounts of iron that had accumulated in the ocean waters (see Chapter 6).

The Great Oxidation Event, which began at about 2.4 billion years ago, removed yet more dissolved iron from the sea, but this was not so much by simple rust formation. A more complex mechanism swung into action—one mediated by sulphur. The oxidation of the land surface also oxidized sulphur compounds in the rocks, and for the first time sulphate ions began to be washed into, and accumulate within, the ocean waters.

At the surface of those oceans, where the water was oxygen-rich, the sulphates remained as they were. However, in the large volumes of oxygen-starved water below the sulphate ions were converted into sulphide ions by bacteria as these stripped away the oxygen atoms from the sulphur (sulphate is, after oxygen, the energy source of choice for bacteria). The sulphide then combined with dissolved iron to form minute crystals of iron sulphide, otherwise known as the mineral pyrite (which also has the popular title of fool's gold). It is certainly a beautiful mineral, with its golden sheen, and it typically crystallizes in the water column as tiny, elegant, raspberry-shaped clusters of micro-crystals called framboids. These fall in countless numbers on to the sea floor, taking the iron and the sulphur with them.

This particular type of ocean—known as a sulphidic ocean—seems to have been the predominant ocean between about 2.2 and 1.3 billion years ago—that is, for much of the Proterozoic Eon of the

Precambrian. The mass crystallization of sulphides was a very effective way of cleaning iron out of the seawater—and not only iron, but trace elements such as molybdenum, zinc, and copper. All became part of the ocean-wide pyrite rain. All of these elements, too, are essential nutrients to living cells, being vital components of certain complex macromolecules of the cell machinery. And so it has been speculated that sulphidic oceans were also nutrient-starved oceans (no matter how much carbon might have been around). In such a view, this long-lived chemical environment acted as a brake on biological evolution for that billion years before oxygen managed to build up to sufficient levels to seep into the depths of the ocean. When that happened, a little over a billion years ago, sulphate became stable throughout most of the water column, and the throttling of essential trace elements stopped, allowing life to progress once more. A modern ocean, chemically speaking, had been ushered in (see Chapter 6).

Sulphidic oceans have virtually vanished from the world today, with one significant exception: the Black Sea. This almost completely separated arm of the Mediterranean Sea is too isolated and too deep to be affected by the currents that stir oxygen into the bulk of the world ocean. Below the sunlit surface, teeming with fish, most of the 2-kilometre-deep column of water is stagnant and poisonous, being charged with hydrogen sulphide. It is also something of a magnet for trace elements. Despite being only a tiny fraction of the area of the global ocean, the Black Sea may be responsible for as much as half of the total removal of molybdenum from seawater—and thus vividly demonstrates the far-reaching effects of a change in a single chemical parameter, that of the availability of free oxygen.

The saltiness of the ocean, then, has not been a constant of the world ocean, nor something that has simply been building up through geological time, nor a one-dimensional phenomenon that simply reflects the amount of dissolved 'salt'. Rather, it has been a protean phenomenon that has expressed itself through a succession of different chemical patterns and processes, which have waxed and waned over time. These processes were pushed and pulled by tectonics, climate, and biology (and, to reciprocate, they did their fair share of pushing and pulling the climate and biology of the day).

The changing saltiness of the oceans has another function, day-by-day and year-by-year. It helps stir and mix the oceans and drive the great ocean currents. The Earth's elaborate liquid conveyor belt is our next port of call.

5

Moving the Waters

It is hard to get a proper sense of the scale of the oceans. To us they seem just enormously deep. A thousand adult humans could—in theory—form a column on an average part of the ocean floor by standing on each other's shoulders—and yet the person at the top would still not have reached more than half-way towards the sea's surface. But now look at it on a planetary scale. The Atlantic Ocean averages 4 kilometres deep—but is also 4,000 kilometres across. At a thousand times wider than deep, it is therefore much, much thinner, relatively speaking, than a pancake: it is an ultra-thin, curved liquid skin, occupying the faintest of depressions on a planet that is smoother—relatively speaking—than a billiard ball. At this scale, the Earth is a ball of rock with a partially damp surface.

Yet this thin curved skin of water is in continuous movement. And it's a good thing that it is. Returning to our familiar human perspective of seeing oceans as endlessly vast and deep, it is the movement of the water that takes oxygen down to the very bottom of the sea floor and keeps animal communities alive, even in the most profound ocean trenches such as the almost 11-kilometre-deep Marianas Trench in the Pacific Ocean. It is the movement of water, too, that brings nutrients into the most productive fisheries in the world, such as off the coast of Ecuador, where enormous shoals of anchovies feed seabirds and fishermen's families alike, and, bizarrely, in another oceanic hotspot of life right next to the barren Skeleton Coast of Namibia. Deep waters there may be, but they are certainly not still.

What have people known of these vast oceanic movements? They long needed a practical understanding of what it takes to survive on the high seas, for humans have sailed the seas virtually since our species originated, over 100 thousand years ago. By 40,000 years ago they had reached Australia—no mean seafaring feat, even allowing for the lower sea levels of those days. Five thousand years ago there was seagoing trading from the Mediterranean to the Persian Gulf to the Arabian Sea. When the Phoenician empire was in full cry, vessels might—as the accounts of Herodotus hint—have reached the Sargasso Sea. What these ancient peoples knew of, or understood, or wondered about the oceans, we know very little. Before writing became sophisticated and widespread, communication between generations was through the creation and memorization of epic poems, in which fact, exaggeration, drama, and myth were inextricably mixed.

The greatest of the epic poets, and one of the first to have 'his'[62] poems preserved in writing was Homer (*c.* 850 BC), author of the *Iliad* and the *Odyssey*, and whose poems were fixed in text, perhaps even in his lifetime. Here, one gets glimpses of the oceans, as were then understood. Homer portrayed the god Hephaistos making an immortal shield to protect Achilles at Troy, with the rim of the shield symbolizing the ocean as a mighty river encircling the land. This idea of the oceans as a river echoed ancient ideas stretching back to the Minoans, Babylonians, and ancient Egyptians. But what made the oceans move? The Greek poets and philosophers were perplexed. Socrates, as quoted by Plato, envisaged a vast set of subterranean cavities, one of which pierced right through the Earth to come out the other side, through which the Earth's waters thundered and surged, to transmit their energy to Oceanus at the surface. Later, Aristotle dismissed the idea of subterranean caverns, but did not come much closer to establishing how ocean waters moved. Perhaps it was something to do with the way the rivers flowed into them, he speculated.[63]

Aristotle's interpretation of the world, together with the dogma of the Church, dominated—or perhaps stifled—Western scientific enquiry for much of the next millennium. When these philosophical shackles were eventually loosened, it was the practical observations of sailors that yielded clues as to the movement of the ocean waters. For instance, in the first transatlantic journeys it became clear that the journey across from North America to Europe was easier and shorter than the return voyage. Going westwards, the early ships were often pushing against a current that was trying to drive them back home. The journey could take more than two weeks longer in this direction than when coming back from America, although some learned to take other routes to avoid the delaying factor. Then there was the strange way in which the ocean temperature changed from shallow to deep waters. Around the equator the surface waters often reach a comfortable 30 degrees Celsius or so. However, when buckets or large bottles were tossed overboard on the end of a rope, allowed to sink to 100 metres or so, and hauled back, the deeper waters were shown to be much colder, the temperature having plummeted to 10 degrees Celsius or less. These early sailors had found a temperature structure that we now call the thermocline. The first record of this seems to have been made by a slave-trading ship in 1751. The captain, one Henry Ellis, was surprised[64] to find that below the Sun-warmed surface layer of water—even in the tropics—the water is icy cold, and the cold extends right to the ocean floor. He was pleased, though: in that burning climate, it allowed him to chill his wine to an agreeable temperature.

That extraordinary eighteenth-century politician, postmaster, printer, author, inventor, satirist, musician, civic activist, diplomat, *and* scientist Benjamin Franklin—a man with such a wide reach that he was called the most accomplished American of his age—realized from these measurements that something was stirring in those depths. If the ocean was still, the Sun's heat would slowly reach through to the ocean depths, and warm them through to the ocean floor. He realized that, therefore, they could not be still. The cold deep water must have travelled across as a submerged current from the far polar regions.

He pondered the question of the mysteriously variable transatlantic crossing times, too, interrogating experienced ships' captains about the courses they took, about which voyages were fast and which were slow. He found out enough to plot on a map of the Atlantic a mass of water that continually flowed across this ocean from west to east, creating an adverse current of some 5 kilometres an hour to those who tried to sail against it. He called it the Gulf Stream. His map—even though he reprinted it a couple of times—was ignored virtually completely, as was, for a long time, his advice to ships' captains on how best to steer a path around the current. This did not dampen Franklin's innovative spirits in oceanography. Late in his life he published a series of novel ideas on ship design, including a soup bowl designed to keep stable in rough seas. Creature comforts, he knew, were important in a sailor's life.

What, though, was driving the ocean currents that were the object of his curiosity?

Driving the Currents

The prime driver of this movement is, one way or another, the Sun. Some 174 petawatts of energy continually shine down on the Earth. A petawatt is one of those words that give little sense of its meaning, but baldly stated it is 1 million billion watts. That might not get us much further in understanding, but another way of putting it is that it is roughly 50 times contemporary humanity's total energy output. And we humans, one must remember, are no slouches at the production of energy.[65]

Part of the Sun's energy—some 30 per cent—is simply reflected back into space. The rest is absorbed. Some of that gargantuan energy input goes into driving the world's weather systems. The air, warmed by the ocean, rises and carries yet more energy with it in the form of water vapour evaporated from the ocean. As the air rises, it cools. The water vapour condenses as cloud droplets and then rain, releasing that transported energy and driving those weather systems yet faster.

Most of the Sun's heat, though, goes into the land and—especially—the oceans. The oceans, indeed, absorb about four times as much solar energy as do the continents, even though they only take up about three times as much of the Earth's area (the difference is because the oceans are darker than the continents, and so are better at absorbing heat).

The heat translates into oceanic motion in different ways. The most obvious effect, to us as observers, is secondary, through interaction with moving masses of air in the atmosphere. The Earth's atmosphere, heated directly by sunlight and warmed also from below by the Sun-warmed land or sea surface, expands and rises in the tropics, to cool in the upper atmosphere and then descend at mid-latitudes, in Hadley cells. Another couple of pairs of such cells of rising and falling air masses form around the mid-latitudes and polar regions. These moving atmospheric masses help redistribute heat from the hot equatorial regions to the polar regions, where the Sun's rays, striking at such a low angle, do little to heat the frigid surface.

The masses of air, as they move north or south, are twisted sideways by yet another force, one imparted from the Earth's spin.

The speed of the Earth's rotation is something that we are normally wholly unconscious of, but it is nevertheless enormous: a little over 1,600 kilometres an hour at the equator, faster than a commercial jet plane. As this speed of planetary rotation diminishes to zero at the poles, anything (for example wind or water currents, an aeroplane, a bullet) travelling from higher- to lower-speed regions will 'carry' some of its original spin momentum with it—and vice versa, of course, if moving in the opposite direction—and be deflected sideways. This is the Coriolis effect, named after the French mathematician Gaspard-Gustave de Coriolis (1792–1843), who analysed the forces involved, not on the Earth, but on rotating water wheels (he was also to write a treatise on colliding spheres, which became much celebrated among billiards players of a more analytical persuasion).

The end result of the Sun's heat and the Earth's spin on winds is the large-scale cells of atmospheric circulation (Fig. 7), which sailors experience as more or less predictable wind systems such as the 'trade winds' and the mid-latitude 'westerlies'. Superimposed on these large-scale systems are smaller-scale, intermittent weather systems of cyclones and anticyclones, their characteristic rotary motion also being a result of the Coriolis effect. The blowing winds tug on the ocean waters beneath, and so drag the surface layers of water along with them. The trade winds, for example, drive currents of water westwards along the equator, with an equatorial counter-current of returning water to balance the equation.

There are other, more direct means by which the Sun's energy can make the ocean waters circulate. Subtropical waters, for instance, are both strongly warmed by the Sun and swept by hot, dry winds (that are on the descending limbs of the Hadley cells). Evaporation of water in these regions leaves the surface ocean layers richer in salt, and hence denser. If sufficiently dense, this surface water can sink to lower

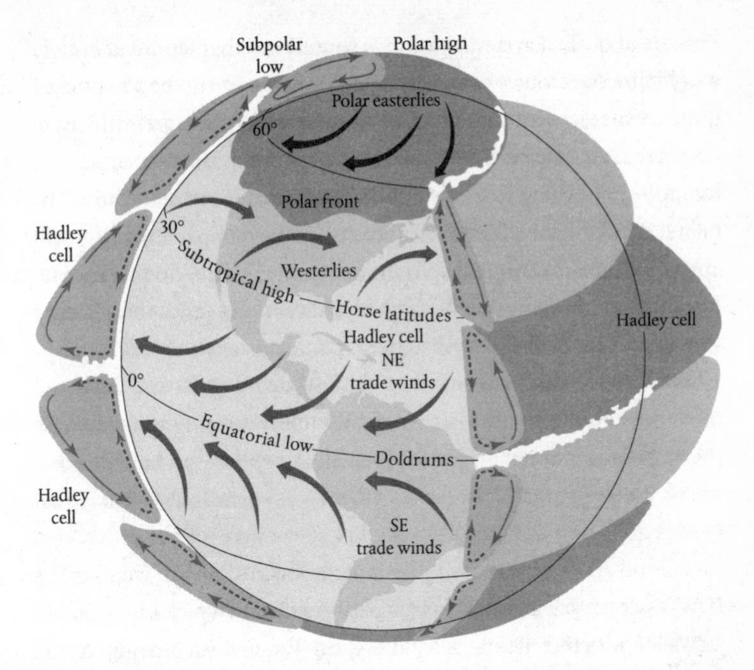

FIG. 7. Atmospheric circulation on planet Earth. The Sun's heat forces the air to rise in the tropics, but it is deflected by the Earth's spin to form a clockwise pattern in the northern hemisphere and an anticlockwise pattern in the south. At mid and higher latitudes are further circulation cells, caused by cold polar air spreading south and intersecting with warmer air from the south.

levels of the ocean or even to the sea floor, and if such sinking is sustained it can generate a continuous current of water deep within the ocean. Such thermohaline currents formed in hot, drying conditions dominated the Earth's oceanic circulation system at some times in its history, as we will see.

However, over the last 30 million years or so an additional means of generating an ocean current has dominated the world's ocean currents. The Earth's global climate since that time has been in an 'icehouse' state, with major ice sheets around one or both of the world's polar regions. Each year the cold surrounding seas form a carapace of sea ice that crystallizes from the surface ocean waters. The sea ice itself is pure water ice, without salt. The water left out of the freezing process therefore contains the excluded salt: that water is dense because of its added salt content, and also because it is cold—and so it sinks. The tail ends of originally warm currents (such as the Gulf Stream) might enter these polar regions too. These are saltier because so much water has evaporated from them on their journey. As these dense, salty waters cool in the frigid polar regions they, too, begin to sink.

These descending masses of cold, salty, dense water are today the main powerhouses of the Earth's oceanic circulatory system. Much of the world's deep water originated from the fringes of the Antarctic, but a classic example lies in the North Atlantic. It is called the North Atlantic Deep Water, and is so often cited in discussions of ocean currents that it has its own acronym, the straightforward but uneuphonious NADW. From its origins, the NADW flows around the southern end of Greenland, before being pushed against the eastern margin of the Americas, both north and south, by the Coriolis effect.

It is quite a current. Oceanographers describe the scales of these flows in units called sverdrups, named after the pioneering oceanographer Harald Sverdrup. A sverdrup is a million cubic metres a second, and the NADW flow approximates to 10 sverdrups, which is more than five times greater than the total flow of all of the world's rivers. It flows along, rather than down, the lower part of the continental slope and the adjoining part of the deep ocean floor as a contour current (because it flows along the contours, rather than across them). Contour currents are usually quite gentle, moving at a few centimetres a second, but now and again there are dramatic increases in current velocity up to, and sometimes exceeding, a metre a second during 'benthic storms'—some of which are perturbations from surface storms that are transmitted down into deep water.

Currents during benthic storms are quite sufficient to erode the sea floor and carry and deposit sediment (and could, we suspect, be enough to knock off their feet any deep-sea diver incautious enough to venture into one). Over the millions of years that these currents have operated they have built up masses of sediment hundreds of metres thick. Geologists call these deposits contourite drifts, and recognized them as thick sediment blankets on the lower parts of continental slopes long before they managed to get their hands on the sediments themselves by drilling down into them. They used to think that contourites would retain a distinct layering that would reflect successive benthic storms, but the borehole cores showed that, between storms, the sea floor creatures so burrowed through and chewed up the sea floor sediments that the original stratification was entirely lost. (Contourites, because of this thorough biological reworking, are difficult to recognize in ancient rock strata.)

This deep current continues southwards, reaching Antarctica, swinging round the continent into the Pacific where it journeys north, then surfaces and heads back, now as a warm surface current, through the Indian Ocean and into the Atlantic. It is not, of course, a simple loop—there are bifurcations, junctions, partings—but in outline it can be so represented. It has been termed the 'great ocean conveyor belt', and it controls much that is in the air as well as the sea, for it is a prime influence on climate (Fig. 8).

The final leg of the journey starts around the Gulf of Mexico, not far off the tip of Florida. It is hot—usually too hot to be pleasant for humans in the summer. The surface water here is also warm, having absorbed energy from the Sun's rays. The whole surface layer is less dense than the deeper waters beneath, and will not sink down into them. The buoyancy caused by this warmth will keep this water near the ocean surface for the next 8,000 kilometres.

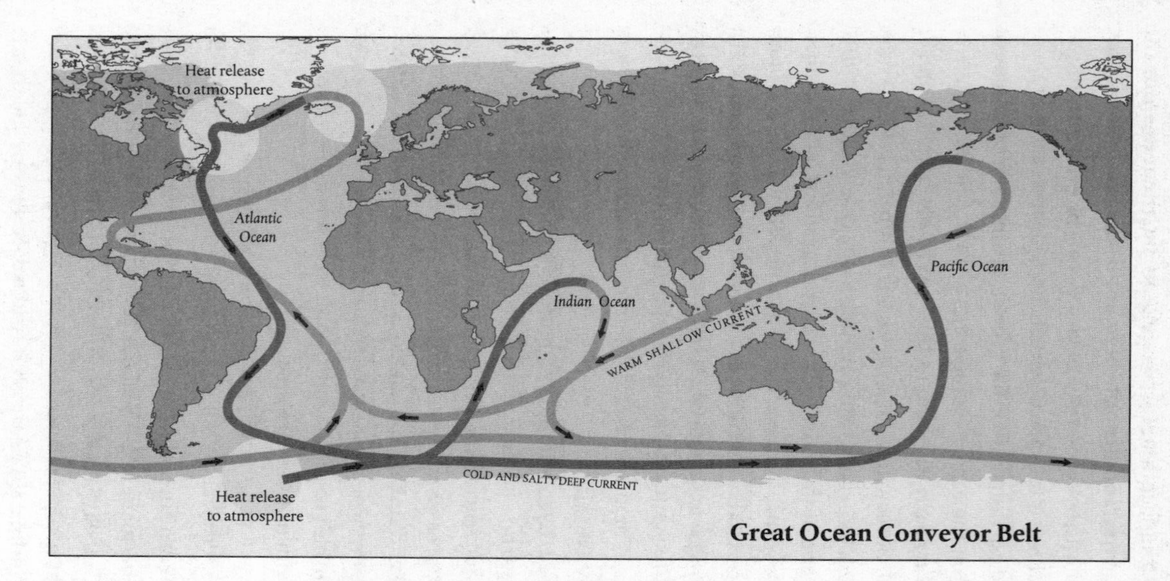

FIG. 8. The ocean conveyor begins in the North Atlantic, where warm water from the Gulf Stream heats the atmosphere in the cold northern latitudes. This loss of heat to the atmosphere makes the water cooler and denser, causing it to sink to the bottom of the ocean. This cold bottom water flows south of the equator all the way down to Antarctica. Eventually, the waters are able to rise to the surface again, continuing the conveyor belt that encircles the globe.

The mass of surface water is moving. Partly, the water mass is pushed by the arrival of other water masses, mostly from off the coast of Africa. Much later on in the journey it will be pulled from the north. Mainly, here, it is being dragged by winds.

The winds first keep the water pushed against the coastline of Florida and then the eastern seaboard of the US. The water here is moving north at nearly 10 kilometres per hour—about twice human walking speed. Somewhere off Cape Hatteras it begins to cross the ocean, as the winds begin to blow mainly to the east. At this point it forms part of one of the most spectacular migrations on the planet. The water here forms a river about a hundred kilometres wide and a kilometre deep, carrying some 150 million cubic metres of water per second—that is 150 sverdrups—on a journey across the Atlantic. This is the Gulf Stream, essentially as mapped and codified by Benjamin Franklin. That water is, incidentally, carrying with it enough heat to supply current human energy needs 100 times over, if it could be tapped. Even without active exploitation, it keeps north-west Europe warm.

Once the Gulf Stream has arrived in the seas around Iceland and Greenland, its heat given up to warm these cold northerly lands, it begins to chill and sink—and so the great journey, where we started, begins once more. For now, this great current system is a constant part of the Earth's geophysiology. It was not always so—and it will likely not be with us forever in the future.

Rerouting Ocean Currents

The world's ocean currents—far larger than any rivers—seem solidly fixed, as permanent as the landmasses that adjoin them. Take the mighty Gulf Stream for instance (Fig. 9). The heat it transports—almost 1.5 petawatts—dwarfs that produced by human activity, and ultimately helps keep north-west Europe pleasantly

FIG. 9. NASA reconstruction of the Gulf Stream and eddies coming off the main currents. Warm water from the Gulf of Mexico flows along the eastern coast of North America before it diverts across the North Atlantic at Cape Hatteras.

warm. Something that large must be a permanent part of our planet's machinery, surely?

It turns out that switching an ocean current on or off, or rerouting it, is geologically commonplace. One way to do it is with truly geological slowness—for instance, as oceans change their shape and continents shift, come into contact with each other, and move apart. The coming together of the Americas, with the rise above sea level of the Panama isthmus, is a classic example. This growing strip of land throttled back, and then cut off, the broadly circumequatorial current that used to flow between the north and south Americas, with consequences that might—via a chain of knock-on effects—have encouraged ice to grow in the far north, and hence plunge the world into an ice age.

Another means is more subtle, but just as effective—and often far quicker. One can keep the geometrical shape of the ocean basins the same, but alter the composition of the water within them. This has been a recurrent pattern in the North Atlantic, and its history is written in ice.

The ice layers that, year by year, have covered Greenland with its 3-kilometre-thick carapace of ice have preserved within them a finely detailed history of the climate of the North Atlantic region, which may be read from the chemical changes in the water molecules that make up the ice layers. Reading this record from ice cores drilled from the Greenland ice cap has shown that, in the approximately 100,000-year-long span of the last great glaciation, the temperature seemed to flicker up and down about every 15,000 years. It took a little while for the significance of this pattern to sink in, and for it to be corroborated with the patchier record of climate change preserved in ocean floor sediments and lake muds. The temperature change in these flickers was not quite as great as that between full glacial and interglacial phases. But it was quite enough to make vivid, life-or-death differences to the early humans then hunting and gathering in Europe. The climate swung from modestly warm to bitterly cold and back again, each 50 human generations or so. The changes, too, could be *fast*. That is clear from the transition that we know best: the most recent one.

The latest of these abrupt swings between cold and warm marks the end of the last great glaciation and the beginning of our warm epoch, the Holocene, and it took place 11,600 years ago. The cold phase is called the Younger Dryas, after a mountain plant called *Dryas octopetala*, more commonly named the white dryad. It has pretty little white flowers, which are unusual in its family (the Rosaceae) for having eight petals. When the weather gets cold the white dryad descends from the mountains on to the plains, and its pollen spreads far and wide (to be detected in layers of pond mud, much later, by scientists).

How long did it take to go from the bitter cold of the Young Dryas to the warmth of the Holocene? This can only be worked out by looking ever more closely at the strata, whether of ice or of mud, that captures this change. The warming used to be thought to be encompassed in a couple of decades—within a human lifetime. One of the latest studies now narrows this change to just *three years*. It was a hemispheric climate revolution, and it was just like switching a radiator on. Literally. Over 1,000 years before, to trigger the beginning of Younger Dryas times, the radiator appears to have switched off just as abruptly. What happened?

The great oceanographer Wallace 'Wally' Broecker has spent a good deal of his career puzzling this problem through. The obvious radiator in that part of the world is the Gulf Stream. Stop that flowing somehow, and those 1.3 petawatts of heat it carries would also be stopped. But the Gulf Stream is *enormous*, as we have seen: how can you block such a thing? No dam could hold it back, and its immediate source, in the Gulf of Mexico, lies close to the tropics where temperature changes in the Ice Ages were small. However, the Gulf Stream is part of a circulatory system that encircles the Earth—so there may be other places where a spanner might be tossed into the works to disrupt its mechanism. Wally Broecker reckoned that the roots of this system, in the far north, might be vulnerable.

This is where the salty Atlantic water is chilled so much that it becomes sufficiently dense to sink, the downpouring water forming the North Atlantic Deep Water, at one major beginning of the oceanic conveyor belt. If the Atlantic brine becomes less salty, though, then the water will not sink, no matter how cold it gets. And if that happens, this part of the global oceanic belt can stop.

Making the Atlantic less salty on regular occasions turned out to be commonplace. The great North American ice sheets, as they waxed and waned, formed barriers to hold back great lakes of meltwater (larger by far than the current Great Lakes) on the North American continent. At times the ice barriers broke down, and the lake water, suddenly released, flowed in brief but biblical-scale torrents that scoured the landscape into deep gully systems and then gushed out into the North Atlantic. Sufficient amounts could be released in a single flood to raise the level of the global ocean by as much as a metre. The meltwater, being fresh, was light and so formed a low-density lid over the North Atlantic. Being light it could not sink, even when very cold, and so North Atlantic Deep Water formation could be suddenly halted. This cut off the Gulf Stream at its roots and brought it to a halt, spreading a chill across the surrounding land. The chill would not be removed until this part of the ocean conveyor belt could—once the salinity was restored—suddenly lurch back into action to spread warmth once more around the north.

This oceanic switch is a remarkable part of the Earth machine, and Wally Broecker's ideas—long fiercely contested—are now widely accepted.[66] This mechanism shows how easy it is to click enormous parts of the Earth machine into different configurations, where patterns of heat and cold, of rainy areas and dry areas, can be more or less instantly readjusted. There really are tipping points in the Earth system: the evidence in the strata is clear, and the strata do not lie (they might be coy about revealing the truth, but that's another matter entirely). When these tipping points are crossed they can transform entire regions, of both land and sea: as the sea is such an enormous store of heat—it can hold much more than the atmosphere—any changes to it also immediately affect the surrounding landscapes.

Could the system be brought to a halt again, to plunge northern Europe into another deep freeze even as the rest of the world warms? Well, there are no more great masses of ice on the North American continent, but there is still a great ice cap on Greenland, and rivers flowing into the Arctic from Siberia. Lately, the Arctic Ocean has begun to freshen slightly as rivers have tended to flow more strongly and Greenland ice has begun to melt at its edges. For the time being this is not enough to disrupt global current patterns. It is a system to watch very closely, though, in coming decades.

But mighty currents do not make up the whole ocean. Over most of the ocean, away from the conveyor belt of currents, the waters are more still. It is time to visit the great oceanic gyres.

Famine and Feast

The Sargasso Sea is a thing of legend. A place where Spanish galleons could be trapped amid the entangling masses of seaweed, and where descendants of conquistadores might still do battle with sea monsters. A place that might hide a sunken continent. A place where nefarious powers might make hideous experiments to threaten humanity, and where superheroes might retire to regain their fantastic powers. A place used, more seriously, as a metaphor to examine how people can also become becalmed within the wide ocean of indifferent humanity, as in Jean Rhys's *Wide Sargasso Sea*.

What is the Sargasso Sea in reality? Travel across it, and more often than not you will see fine weather and a remarkably blue and clear sea—looking down you can often peer through the water for 50 metres or more. There is a great deal of seaweed there, of the genus *Sargassum*, the only seaweed that is adapted to a wholly floating existence. It provides shelter for everything from eels (migrating from both North America and Europe) to turtles, to whales and sharks. These days, the Sargasso Sea is also home to an extraordinary variety of floating plastic (but more on that development in Chapter 7).

The seaweed, contrary to myth, is never thick enough to trap entire galleons. Sailing ships could nevertheless become becalmed there, for the area lies within the 'horse latitudes', those generally high-pressure areas of blue skies and gentle winds that lie outside the atmospheric superhighways such as where the trade winds blow. The horse latitudes, by the by, seem to be so termed because seamen of the old sailing ships were often indebted when they set sail, and so were in what they called 'dead horse time'; they had often paid off their debts when reaching these latitudes, and then traditionally paraded a straw horse effigy on deck—then threw it overboard—to celebrate.

Partly because the winds are calm, the waters are too. The Sargasso Sea is the best-known part of a wider stretch of more or less becalmed water called the North Atlantic Gyre. Fast-moving oceanic currents encircle the Gyre: to the west and north, the Gulf Stream and its continuation the North Atlantic Drift, flowing eastwards and northwards; and to the east the Canary Current, flowing southwards to join the North Equatorial Current south of the Gyre and flowing westwards, back into the source of the Gulf Stream. Held within this circulatory system, the Gyre itself very slowly rotates, under the influence of the ever-present Coriolis effect. There are other gyres: two each, north and south, in the Atlantic and Pacific oceans respectively, and one in the Indian Ocean, each also surrounded by faster-moving currents.

This gyre-plus-current combination forms a complex system, which has been more clearly revealed by chemical clues such as analyses of salt content than by physical trackers such as floating beacons. The floating markers, infuriatingly, did not move in the orderly pattern predicted, but rather took off in myriad mutually inconsistent and incomprehensible directions. What was happening? The arrival of satellite images showed why. On these images, near-circular 'mesoscale eddies' can clearly be seen, 100 kilometres or so across, continually spinning off from the boundary between the slow-moving gyres and the speeding current systems that encircle them. The rotating eddies persist for a year or two, slowing, before merging with the mass of calm water in the gyre.

For the most part the great oceanic gyres are regions of low nutrients and low biological productivity (the Sargasso Sea is a bit of an exception in this). They are far from land-based, river-fed nutrient sources, and in such calm waters it is hard to stir nutrients from the deep waters below the thermocline. Here, therefore, there are thin pickings for every kind of planktonic organism, from viruses to whales. Gyres are shunned by the kind of plankton that need a good nutrient source, such as diatoms and radiolaria, and inhabited by those that can make do with less, such as foraminifers and coccolithophores. It is the skeletons of these hardier organisms that go on to make up the oozes that accumulate far down on the deep ocean floor. These gyres are the oceanic deserts of the world, and they are enormous.

The oceanic oases where marine life blooms, by contrast, are much smaller. To find these, one needs to find a very specific kind of current—one that comes from the deep.

We met Alexander von Humboldt in Chapter 3, being perplexed by the oceans. Nevertheless, he was the man Darwin regarded as the greatest scientific traveller who ever lived, and Thomas Jefferson thought him the most important scientist he had ever met (although Napoleon Bonaparte reputedly told Humboldt, 'You are studying Botanics? Just like my wife!'). Humboldt's reputation survived Napoleon's commentary, and his name now adorns a whole array of earthly and unearthly objects. There is Pico Humboldt in Venezuela, almost 4.5 kilometres high, and, in the same country, plumbing the depths, the Sima Humboldt, a sinkhole. Nevada goes three better to name after him a river, a lake (now dry), not one but two mountain ranges (eastern and western), and a whole county. There is a Mare Humboldtianum on the Moon, and even farther there is 54 Alexandra, the first asteroid to be named after a man, orbiting the Sun. The biological tributes range from Humboldt's lily, *Lilium humboldtii*, to *Conepatus humboldtii*, Humboldt's hog-nosed skunk. And there are many more monuments to him. Of all of these, though, one stands out. Pride of place must go to the Humboldt Current, which flows southwards past South America. It is the most productive part of the world ocean.

The water within it is chilly. Humboldt, on his travels, himself had measured it to be a full 7 degrees Celsius colder than water that was farther offshore. Did it, he thought, therefore come from the south? Indeed it does, and the cold water sucks moisture out of the air—much as a cold windowpane does—to produce frequent mists and fogs. As far as water goes, therefore, it receives rather than gives. Hence, the adjoining coast is one of the driest on the planet, where the Atacama Desert adjoins the sea.

The wind there blows offshore. It helps, thus, to combine the life-giving sunlight in the surface waters with the nutrients stored in the ocean below the thermocline, at depth. The wind pulls the surface water away from the coastline, and deep, nutrient-rich water wells up to replace it. This triggers an explosion of planktonic algae, fed upon voraciously by swarming zooplankton, in turn devoured by fish. Four species, in particular, run amok: anchovy, sardine, jack mackerel, and chub pollock. Just in that small space—less than 1 per cent of the world's ocean surface—they can make up to one-fifth of the world's total fish catch.

These extraordinary fish populations have been a source of food long before humans arrived on the scene. Larger fish and dolphins joined in the feast, and so did seabirds—so prodigiously that the excrement from their feasting could pile more than 20 metres thick on the coast and adjacent islands. It was the perceptive Humboldt, too, who recognized the significance of these guano deposits, and his writings sparked off a large industry in the nineteenth century, shipping this nitrogen- and phosphorus-charged stuff to the fertilizer-starved farmers of Europe. Humboldt, a humanist to the core, would have wept to see some of the human consequences of this industry, which fed booming European populations with the oceans' bounty. It was vile, arduous work to dig out the stinking guano (that had compacted to rock-like consistency) in the infernal desert heat. The imported Chinese labourers had to shift 100 barrow-loads a day, and those too weak to continue had to work on their knees to pick stones out from the guano. Small wonder that guards had to be posted, to stop them from leaping into the sea to their death to escape this hell on Earth.

The booming fish populations are not constant, though. In El Niño years the dynamics of the Pacific Ocean change, the winds slacken, and the supply of nutrients from the deep diminishes. The fish populations crash, and so do those of all the creatures dependent on them, the larger fish and seabirds. The human fishers of anchovy and sardine go through lean times too—and the effects reverberate through the world in altered weather patterns that can cause droughts in Africa and floods in California and in the Atacama Desert. It is a reminder of how sensitive the world is to minor fluctuations in the behaviour of the oceans.

However, there are steadier and more reliable motive forces to the movement of the waters of the seas. These too, though, have excited awe and terror in human hearts. Their power comes from the heavens.

Moon Madness

The poets and philosophers of ancient Greece had no means of grasping the nature of ocean circulation, or even of knowing that the waters truly did circulate. But they could appreciate the surface drama of the oceans. From that, they manufactured an image that then fed into human fears for two millennia and more. Homer wrote of Odysseus's year-long sojourn on the island of Aeaea, and of his relations—part opponent, part lover—with its local sorceress, Circe. Odysseus escaped by steering between the twin perils of Scylla and Charybdis, monsters that inhabited—or personified—a high cliff and a whirlpool respectively.

Aeaea seems, alas, to have been a place as lacking in reality as it is in consonants. But Scylla and Charybdis had more reality, and it is generally agreed that they were located in the Straits of Messina between mainland Italy and Sicily (and so not far from the site of a previous Mediterranean catastrophe that was played out in emphatically pre-Homeric times, as we related in Chapter 4). Charybdis, in Homer's prose, is a fearsome thing, drawing the sea's waters down into a vertiginous vortex (at the bottom of which an observer's terrified gaze might see the sea floor, exposed) before dashing them out again to spray high into the air. Even the mighty Odysseus only lived through the encounter by clinging to an overhanging fig tree while his raft was dragged down into Charybdis's maw. Charybdis is the model for other whirlpools, such as the Norwegian Maelström, the setting for one of Edgar Allen Poe's early spine-chillers. He wrote of an encounter with this deadly funnel of spinning water that turned a bold youthful sailor, overnight, into a white-haired, unhinged wreck of a human. The Maelström, too, vanquished Captain Nemo's mighty submarine *Nautilus* in Jules Verne's *20,000 Leagues Under the Sea*.

Charybdis—and even the Maelström—have been overhyped. Neither, in reality, are the kind of ship-devouring undersea tornado described in the fictionalized accounts. Nevertheless, strong currents do run in those and other places, and these undoubtedly alarmed the early sailors in their primitive boats. These are concentrations of a kind of power that many ancient seafarers were well aware of. It is the power of the tides.[67]

Tides seem to be rather contradictory phenomena. On the one hand, they are so predictable in their operation that today tide tables can be printed years ahead, to allow seaside holidaymakers to plan their shoreline walks carefully so as not to be cut off by the rising waters. On the other hand, the distribution of tides seems entirely capricious. They are generally a feature of shallow seas around continents, but in some places they are absent (the Mediterranean is *almost* entirely tideless, for instance) while elsewhere they are simply enormous: in the Bay of Fundy, Canada, the difference between high and low water is a staggering 16 metres, and not far behind is the Severn Estuary in England with a tidal range of up to 14 metres (yet go round the UK coastline to, say, Swanage and the tidal range goes down to a couple of metres).

Where tides exist, though, their mechanism is so predictable that a connection with the phases of the Moon has been made since the days of the ancients (who were good practical astronomers). In everyday understanding, this is often simplified to say that high tides represent the effect of the pull of the Moon's gravity upon the Earth's waters. This is only partly true, for there is simultaneously a high tide too on the side of the Earth *opposite* to the Moon, as well as on the side facing it. How so?

This reflects the delicately balanced Earth–Moon system. The Earth and Moon are held together by gravity, true, but if that was all there was too it, then the Moon would *very* soon come crashing down on to the Earth. The Moon and the Earth are kept apart by another force, exactly matching the gravitational pull—and that is the centrifugal force between these two bodies as they whirl rapidly around each other; without gravity, this force would make the Earth and Moon instantly fly apart.

It is the difference between the two forces that raises the tides, and the Earth is in effect rotating beneath an envelope of water that is shaped into two gigantic bulges. From the point of a human observer the planet appears stationary, so it appears that a great, slow-moving wave of water passes by twice a day, as the tide goes in and out. The tides can be amplified or diminished at any place on Earth depending mostly on the shape of the sea floor, which can either act to resonate with the tides and hence make them larger (as in the Bay of Fundy) or dissipate them.

The Mediterranean is mostly the wrong shape to generate strong tides. In Homer's day, as now, the typical difference between low and high tide is about 20 centimetres. On a wave-stirred surface, that is too small to be easily noticeable, and Homer may not have known of tides at all, let alone of their connection with the Moon. Charybdis is an exception. It lies at the site of a narrow, deep channel (the Strait of Messina) between landmasses and, by a freak of submarine geography, the timing of low and high water at either side is offset by several hours, so that the water slopes first in one direction along the Strait, and then in the other. It's not much of a slope in sea level—about 5 centimetres over 3 kilometres—but this is nevertheless enough to generate currents of up to 2 metres a second when focused along this narrow waterway. When those currents begin to reverse as the slope on the surface of the water see-saws once more, vortices—whirlpools—are set up. These, particularly when appearing seemingly out of nowhere on a calm summer's day, would have looked horrifying and mysterious to ancient Greek sailors.[68] The original Maelström is a similar phenomenon but, with stronger tidal currents to start with, genuinely more fearsome.

Tides, like wind-formed waves, mainly act to stir the surface and near-shore waters of the Earth (out in the oceans the tide is small—usually less than 30 centimetres separates high water from low water). The effects are usually greatest in estuaries and embayments that funnel and concentrate the tidal forces, but here and there they can also reach down to several hundred metres—for instance into deep submarine canyons, some of which regularly have their bottoms swept by tidal currents.

Earth has regular but modest tides, from a small Moon (a little less than one-eightieth of the mass of the Earth) in a more or less circular orbit. In the past, it used to have more rapidly repeating, higher tides from a Moon that was closer (and this can be read, remarkably, by careful analysis of ancient tide-influenced strata). Venus, with no moon of its own, and Mars, with a couple of pathetically small orbiting rocks barely meriting being called moons, will have never had tides other than those generated from the distant Sun. These planets do not now possess oceans to be tide-swept (although they may have had them in the past, as we discuss in Chapter 10). Tides elsewhere in the solar system can be substantially fiercer than on Earth: Io, a moon of Jupiter, is so strongly kneaded by being held in the grip of that giant planet that it is melting inside, and volcanoes are continually erupting. Now that we can detect other planetary systems far out in space, we find these to be remarkably diverse, and often marked by tight, bizarrely looping planetary orbits. Tides will be a key mechanism for governing the behaviour of many far-off ocean-bearing worlds.

Waves

We feel the movement of the air as winds, and these winds drag on the ocean surface, helping to drive its current systems. But the most obvious visible effect of winds is to stir up waves on the water surface. Once generated, waves can then travel thousands of kilometres across the ocean to crash on some distant shore. It's a fabulously efficient transport system—but it is *energy* that is being moved such distances, not water. The movement of the water here is strictly local, the water particles simply moving round and round to create the wave-form, as one can see by sitting on a beach and idly watching the bobbing of a piece of driftwood amid the passing waves. This movement dissipates downwards although, even with the biggest waves generated in the greatest storms, sea floors more than about 150 metres deep will experience little or nothing of the seething turmoil at the water's surface.

On a beach holiday, it is fun to see how small waves work by paddling out to sea a little and looking down at the shallow sea floor around one's toes (or, if you want to be properly scientific about this and get a really good view, you can snorkel and look downwards). On the sea floor, the passing waves create a kind of backwards-and-forwards motion, which piles the sand into little symmetrical, sharp-crested ripples. These 'wave ripples' are quite distinctive. When found preserved in ancient strata, wave ripples are a telltale clue to shallow water conditions on the edge of a sea or a lake.

Most of the energy carried within waves is finally expended on the world's shorelines, and here even the casual observer, carefully conserving their own energy in a nicely positioned deckchair, can observe the geology at work. Beaches are built by swash and backwash and longshore drift, transporting, smoothing, and sorting sand and pebbles as the waves break on the shore. Beaches can be extended far out along remarkably straight spits, trapping lagoons behind them. Cliff lines and wave-cut platforms are carved by the explosive crashing of waves on to rocky headlands. These are obvious features of classical geomorphology, and they would likely form on any planet with a combination of atmosphere, land, and ocean.

If the storm winds blow long enough, though, the water itself does begin to move as a mass as it is dragged by the winds. Beneath a major storm system, water may be piled up on to land by driving winds, a process often amplified by low pressure sucking the water upwards. The effect can be devastating, as when Hurricane Katrina overwhelmed New Orleans in the late August of 2005, and Hurricane Sandy brought destruction to New York in the winter of 2012. When the winds die and the barometer rises, the water floods back to sea.

A storm ebb surge, as it is called, takes with it mud and sand, exhumed shells, and debris scoured off the land by the storm waves. Today that can include fragments of masonry from buildings swept away, and metal and glass fragments from storm-swept vehicles. All this is spread over hundreds or thousands of square kilometres of sea floor, to form, in a turbulent few hours, a new stratum on the seabed that, near land, may be more than a metre thick. Geologists call these storm layers, or tempestites—a Shakespearean resonance quite in keeping with the drama involved.

Farther out to sea, in deeper water, small fair-weather waves do not reach down to the sea floor. This is the realm of the storm waves that, in rough weather, create a different and rather puzzling structure. Not quite a set of horizontal layers, and not quite the steeply inclined laminae of fossilized dunes—rather, packets of layers, usually a few tens of centimetres thick, which seem to be more subtly wavy, to pinch and swell, often in rhythmic sets. When the top of such stratal surfaces are exposed, with a bird's-eye view they look a little like the top of an enormously magnified golf ball, showing rhythmically arranged gentle hummocks a metre or so across and a few centimetres high. Nothing spectacular—and easy to miss if you didn't have your eyes wide open. Peer at the bottom layers of these subtly dimpled units, though, and they are not as gentle as they look. They are often full of smashed and broken seashells.

Hummocky cross-stratification, this type of layering was termed, and it turned out to be so common as to be given its own acronym: HCS (geologists, alas, have a predilection for this kind of abbreviation). What formed it? This type of layering is now most convincingly interpreted as the results of great offshore storms. HCS should also be a planetary constant. It should be high on the checklist of clues to be sought by the Mars rovers as they trundle across the strata of the red planet. If they do ever find HCS on their travels, then that acronym will become headline news worldwide: a sure-fire indication that this now freeze-dried planet had, once upon a time, deep and stormy seas.

Living Oceans

The oceans of an icehouse Earth circulate energetically. On a world with abundant polar ice there is never a shortage of freezing winds to chill water to its sinking point, or that armour-plating of sea ice, crystallizing from the frigid waters, to add to the continuous downdraught of dense brines. Sweeping downwards from each pole, the interlacing loops of the Earth's ocean current system provide a supply of cold and—crucially—oxygen-rich water to all parts of the world ocean. They allow multicellular life forms to thrive almost everywhere, living suspended in the water or on the sea floor, even in pitch darkness at great depths and crushing pressures.

The deep life of the oceans came as something of a shock to the world of nineteenth-century science. The persuasiveness and charm of one of the greatest—and nicest—naturalists ever to have lived had quite a lot to do with this. Edward Forbes was born on the Isle of Man, and on that island he developed an interest in animals and plants that never left him, even when he should have been moving on to the more serious business of earning a living. As a young man he might have been called a wastrel, having a talent for art (he was a splendid caricaturist) but never developing it professionally. He enrolled as a medical student in Edinburgh, but then cheerfully neglected his studies to go out on impromptu expeditions to collect flowers, insects, and seashells.

In his case fortune favoured the blithe of spirit. Kept alive by a small allowance from his father, he threw himself into the mid-nineteenth-century world of natural science, and shook it up more than a little. At the 1839 meeting of the British Association in Birmingham, for instance, he shrugged off the formalities of those procedure-bound days by decamping to a local pub, the Red Lion, taking a fair proportion of the attendees with him. There, in between fuelling themselves with beef and beer, and joke and song, they debated the great scientific issues of the day. Forbes's 'Red Lions' became a standard event at subsequent British Association meetings, and some idea of their flavour is given by the way that they did not express their approval or disapproval of scientific points raised by a vote or a show of hands, but—in true leonine fashion—by growling and roaring and fluttering their coat-tails (Forbes's own technique was held up as a model for the younger Lions).

Small wonder that he had a following who loved his style and generosity of spirit. He went on to become the British Geological Survey's first palaeontologist, and was ferociously productive, publishing no less than ten papers in the first ever issue of the Geological Society of London's classic *Quarterly Journal*. But it was his work as a scientist on board a survey ship, HMS *Beacon*, in the Mediterranean in 1841 that allowed him to develop his notion (the 'azoic hypothesis') of deep, cold, lifeless oceans. The idea was eminently logical: after all, life is seen to diminish as one climbs mountains, and the snow-bound tops of the Alps and similar mountain ranges are essentially lifeless. There would be a natural symmetry to a similar trend as one descends into the cold, dark depths of the oceans.

It was not just a grand idea. The data that Forbes painstakingly collected from the dredging buckets as HMS *Beacon* traversed the Mediterranean seemed to back him up. Shallow-water life in that sea flourishes in an abundance of forms. Both the amount and the diversity of life were then seen to diminish as the ship moved to sample deeper parts of that sea floor. Forbes projected this trend to even deeper, unsampled waters. The deep ocean must be dead, he said.

The hypothesis, propounded with Forbes's characteristic energy and eloquence, took hold and held sway—indeed for far longer than it should have. For the sounding lines of commercial ships, even before the *Beacon*'s expedition, were sporadically dragging up organisms (worms, starfish) from even deeper waters—from 2 kilometres and more down.[69] These should have killed the azoic hypothesis stone dead. As chance had it though, the few early reports of deep-sea life were made by people who were either unreliable of character (they were known to make things up) or who were simply bad-tempered. They simply could not compete with Forbes's (deserved) influence or popularity, so it took some more years of further reports of very deep-sea life to drag the azoic hypothesis down. The collapse of his grand idea didn't do much to harm Edward Forbes's reputation—he is still regarded as the father of marine biology—and the Red Lions, growls and fluttering coat-tails and all, carried his spirit forward after his untimely death at the age of 39.

The scale and diversity of deep-sea life was only revealed in its true splendour when the bathyscaphes made their way down into those depths and began to make systematic observations. In a way, it should not have come as so much of a surprise. There is a steady food supply, with all of the organic particles sinking from the productive surface layers of the ocean. The conditions are stable: it is cold, but never gets below freezing, and the organisms down there are adapted to life at just a few degrees above freezing. The waters are at high pressure (hence the need for bathyscaphes for us surface-dwelling humans), but the organisms living there have the pressure equalized between their tissues and the water outside. It is dark—but those organisms have senses other than sight (and many of them also carry their own light sources). Crucially, there is oxygen, courtesy of the oceanic conveyor belt.

In its way, this life-giving system is just as intricate and as remarkable as the system of arteries, veins, and capillaries that brings oxygen to every cell within our tissues. The oxygen in the oceans has to reach water depths of 5 kilometres and more, across distances of thousands of kilometres. And it has not just to be supplied to living organisms, but to be present in sufficient amounts to help oxidize the constant supply of dead organic matter that is drifting down as 'an eternal snowfall', in Rachel Carson's evocative phrase. Too little oxygen and too much organic matter and, as in an overfertilized lake or ditch, the oxygen is used up, the waters turn anoxic, and they can no longer support multicellular life.

The supply is not quite even. The Atlantic Ocean is a little better ventilated than is the Pacific, and there is a widespread layer at intermediate depths, in-between the most rapidly moving parts of the conveyor belt, that is known as the Oxygen Minimum Zone. But the parts of the ocean floor that, today, have no oxygen are so few that they have become geological classics. Offshore from Santa Barbara in California there is in effect a hole in the sea floor, about 30 kilometres across, called the Santa Barbara Basin. It contains a puddle of submarine water which, protected by the surrounding basin sides, is excluded from the surrounding ocean circulatory system. The water there contains no free oxygen, and so there are no worms, no crustaceans, no starfish to disturb the layers of sediment that drift down, year by year. These layers are a treasure-store for marine scientists, who pull up cores of the finely laminated strata from the sea floor. They can measure within them, for instance, the scales of fish that lived in the waters above, and compare these with fisheries records to pinpoint good years and bad years for fish stocks. They can then go down to layers of sediment from well before the times when humans made records, and reconstruct the climate and oceanography of the sea in fine detail, deep into prehistoric times.

There is another such small basin, the Carioco Basin, offshore from Venezuela. And there is also that much larger but completely landlocked sea, the Black Sea. Below its sunlit, productive surface layers, the great, still bulk of the 2-kilometre-deep waters are so depleted in oxygen that they are not just anoxic but euxinic—they contain dissolved hydrogen sulphide, a gas with the humorous connotations of rotten eggs and stink bombs, but one that is as deadly to most living creatures as is hydrogen cyanide.

Such isolated areas are rare outliers within today's well-ventilated oceans. But such areas were once expanded enormously, to cover large parts of the globe, at times when the oceans' circulatory system worked less well than it does today. All it takes to visit them is the time machine afforded by rock strata and then one can, almost literally, walk on those surreal sea floors of the deep past.

Hothouse Oceans

We are living today on a kind of world that represents a minority of Earth time. Over most of our planet's history the Earth has had little or nothing in the way of polar icecaps. Take, for instance, the Cretaceous Period, around 100 million years ago. The tropics then were somewhat warmer than today—perhaps more than 10 degrees Celsius warmer at times—but the high latitude regions could be as much as 50 degrees warmer. Instead of ice and tundra, there were polar forests inhabited by polar dinosaurs adapted to six months of continuous daylight and six months of darkness. The temperature gradients between pole and equator, therefore, were much lower than at present, and it is still a little mysterious quite how the heat was transported so effectively from low to high latitudes.

On such a world, there is little chance of forming large volumes of cold deep waters by sea ice formation, or by cooling with freezing winds to power the oceanic conveyor belt. These oceans, therefore, must have been driven by forces different to the ones operating today. Quite how did they work?

The great palaeo-oceanographer Bill Hay has spent a good deal of his long and colourful career, which has oscillated between post-war Europe and North America, puzzling over this problem. It was an ocean system which, he observed, is utterly unfathomable if one follows Charles Lyell's dictum that the present is the key to the past. For so much then, besides global ice volume and temperature, was different.

Sea level, for example, might have been more than 200 metres higher than today. Part of this amount would have come from the melting of virtually all the world's ice—but only part (today, melting all global ice would only raise sea level by some 70 metres). Part—maybe a few tens of metres—would have come from thermal expansion of the oceans. And part was likely not related directly to the world's climate, but reflected global tectonic patterns. If the Cretaceous ocean crust was forming faster than it does today, that crust would have been hotter overall than today and therefore, through thermal expansion, would have been raised higher, displacing the overlying ocean water upwards. The increased volcanism would have belched out more carbon dioxide too, helping to warm the Earth's climate through an enhanced greenhouse effect.

The ocean waters would have spilled over the continents, submerging much of their area. This would have blurred the sharp distinction that exists today between the shallow waters of the continental shelves and the deep ocean waters. This distinction can be seen physically as a sharp line (the ocean front) separating clear blue ocean water from the shelf waters that are often turbid and sediment-laden. It is a key boundary, typically separating very different communities of marine organisms.

In the Cretaceous, with its higher sea levels, the ocean front would have disappeared or become severely blurred. In effect, oceanic conditions moved across on to the continents. The iconic chalk strata that formed in late Cretaceous times are in effect a deep-sea deposit—the settling of trillions of coccoliths, complex microscopic skeletons of calcium carbonate made by single-celled planktonic algae—that drape what was previously a shallow sea floor or even land.

With little or no ice at the poles, Bill Hay sees the Earth's climate system as having been held in a less tight grip than it is today.[70] Without a supply of cold polar-derived deep waters, circulation in general would be more sluggish. Each of the polar regions, instead of being permanent mirrors of ice to reflect the Sun's light and heat, would have alternately warmed and cooled in their winters and summers, destabilizing the Hadley cell wind patterns, and without the constant tug of steady wind patterns there would not be a modern-day organized oceanic conveyor belt of deep and surface currents demarcating central gyres.

Instead, there would have been a more sluggish and chaotic pattern of circulation. Some movement of water downwards, to ventilate the deep ocean, might have occurred through the modest formation of dense sinking water masses generated by evaporation in arid climate belts. Perhaps most ventilation was achieved through large storms, which could generate 'mesoscale' eddies, perhaps of the order of a hundred kilometres across or so (and so similar in scale to those that we mentioned earlier in this chapter, spinning off from the boundaries of the oceanic currents to drift into the slowly moving gyres). In the Cretaceous such eddies might have played a crucial role in pumping water, here and there, through the oceanic water column.

This kind of mechanism is less effective as an oxygen provider than is the present-day oceanic conveyor belt that loops through today's oceans, and therefore in the Cretaceous hothouse the oceans were more prone to anoxia than now. At times, this tendency spread widely to suffocate much of the ocean floor. Many successions of chalk strata contain, punctuating the usual dazzling white rocks, layers up to several metres thick of dark grey, carbon-rich chalk. These mark times when, over much of the sea floor, the faltering oxygen supply mechanism broke down completely. Organic matter, drifting from the surface, was preserved and buried (to give the strata their dark colour). Multicellular life vanished from these sea floors. These are called Oceanic Anoxic Events, and individual instances can be traced across much of the world.

The main episodes of this form of ocean suffocation seem to have lasted in the order of 100,000 years. They have been linked to the warmest times within the Cretaceous—spikes of global heat that, in some cases, have been more or less firmly linked to external forces such as rare, prolonged volcanic outbursts that would have temporarily increased carbon dioxide levels in the atmosphere.

There seems to be some kind of pattern here. The warmer the climate, the less efficiently oxygenated are the oceans. In the relatively recent Cretaceous though (only the order of a hundred million years ago), anoxic oceans were relatively rare events and are linked with peaks of extreme warmth. This pattern can be taken even farther back in time, to more than half a billion years ago during the Cambrian, Ordovician, and Silurian periods. In those times, anoxia was a much more frequent occurrence—and could sometimes encroach on to shallow waters. Deep-water anoxia, then, was normal, and not a rare catastrophe. In such oceans life was simply adapted to being confined to shallow water conditions and the surfaces of deep oceans. Getting on to the deep ocean floor was a bonus, only intermittently enjoyed by the animals of those days.

Further back still, in the Precambrian, anoxic deep-sea floors were very much the norm. Clearly, oxygen levels in the oceans climbed only slowly and irregularly, from the initial appearance of free oxygen in the atmosphere 2.5 billion years ago.

The physical and chemical structure of the oceans has been constantly changing, as our watery planet has evolved from youth to maturity. This had profound implications for the life that evolved within it—as we shall see in the next chapter.

6

Life of the Oceans

The Earth was born 4.56 billion years ago amid the violent collision of planetesimals. Sometime in the ensuing 100 million years or so it was reborn, at white heat, in the aftermath of the Moon-forming impact. Only after that could temperatures cool to levels that allowed the first oceans to form—although these were likely intermittently vapourized as impacts continued.

About 3.8 billion years ago the last great flurry of asteroids subsided. From then until now the oceans have been a constant feature of this planet. From about the same time, too, there is evidence for life, preserved as the products of early microbial activity. So rapid was this appearance of life, after the period of heavy asteroid bombardment, that microbes might have evolved sometime in those perilous Hadean times,[71] clinging on in rock fractures deep beneath the surface before emerging into a calmer Archaean oceanic realm.

From this beginning, all life in the oceans and on Earth has evolved in an unbroken chain of increasing diversity. How did the complexity of modern ecosystems arise and how long did this process take? For nearly the first half of Earth history the ocean was devoid of oxygen, although rich in dissolved iron. It teemed with microbes adapted to those conditions. The primitive life in those early seas was already changing (and being changed by) the chemistry of the oceans, as the world's greatest concentrations of iron ores began to settle on to the sea floors. Over time, this chemical and biological evolution led to the myriad forms of life of the modern seas. These organisms, from the tiniest microbe to the blue whale, share a single point of origin.

The story of marine life is very largely the story of life on Earth (Fig. 10). It is easy to forget, from our perspective—that of a late-comer, landlubber species, living amid forests and meadows[72]—that life on land in any practical sense is a late invention on this planet. The invasion on to land took place less than half a billion years ago, while the seas have teemed with living organisms for going on 4 billion years. In understanding where we have come from it is, at heart, the oceans that count.

Life Begins

In seventeenth-century Florence, the physician and naturalist Francesco Redi turned his gaze towards decay and corruption, and discovered one of the secrets of life. Redi undermined a central tenet of the long-held view of the world, stemming from Aristotle, which held that life could form spontaneously from lifeless matter. This idea had been used to explain the appearance of maggots in rotting meat, mice springing forth from hay, and even fossils as failed attempts at life. Redi devised some simple but effective experiments to test for spontaneous generation, and these are among the first scientific experiments that involved the use of a control. Redi took glass jars and placed pieces of fish and meat into these: some of the jars he sealed with muslin. Miraculously, maggots appeared in the open jars, but in the jars that were sealed the contents remained free of infestation. Redi surmised correctly that there were no maggots where flies were unable to enter the jars. Redi, thus, may be regarded as the first true experimental biologist. In his native Tuscany though, he is probably

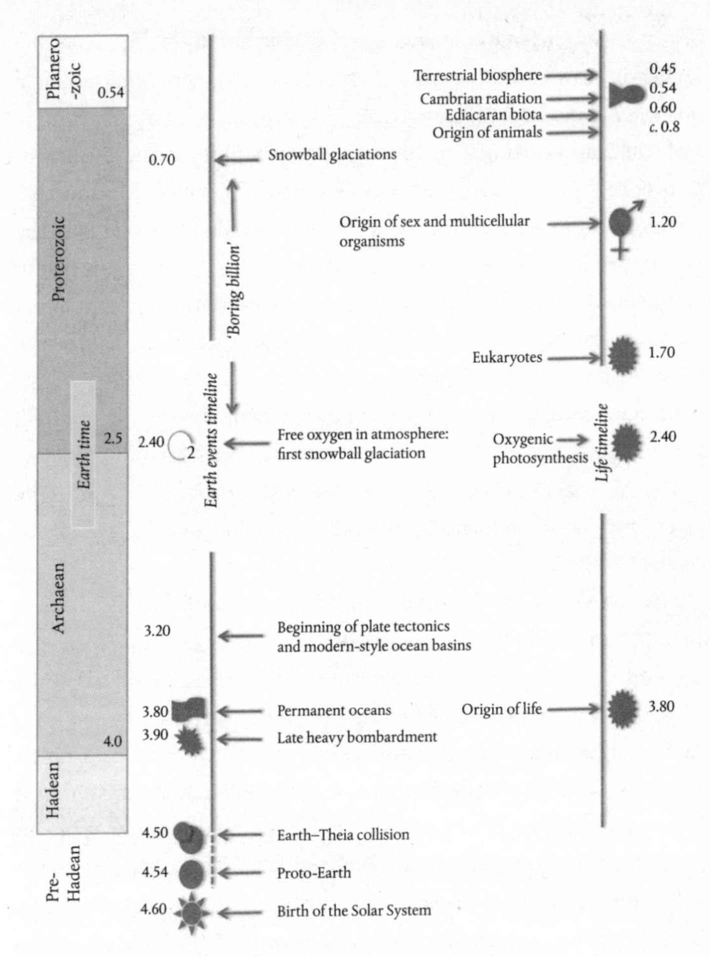

FIG. 10. Major events in the story of life, versus geological time (figures represent billions of years).

better remembered for his poem *Baccho in Toscana*, an ecstatic celebration of the wines of the region.

Despite the more than 300 years that separate us in time from Redi's investigations, some echoes of the idea of spontaneous generation persist today, and have entered our thinking about the origins of life on Earth. How could life emerge from lifeless matter? What possible mix of chemicals (water in particular), energy, and chance could produce the complexity of even the simplest living things? The late British astronomer Fred Hoyle, who was notorious for taking controversial positions on many scientific issues, proposed that life evolving in the oceans of Earth was akin to a tornado ripping through a junkyard with all the disassembled bits of a Boeing 747, and assembling them into a flying Jumbo Jet. In fact Hoyle had his own agenda to push, because he favoured the idea of panspermia: the notion that life arrived on Earth from outer space. His jumbo jet analogy leaned more on the pre-Redi idea of spontaneous generation than on a reasoned scientific analysis of the evidence. In effect, Hoyle was suggesting that life needed to be assembled in one sudden step. The journey to life, though, was not so straightforward.

On Earth there are certain characteristics that apply to the smallest microbe and to the most complex human. Life is self-organized, and is composed of a complicated array of structures and chemicals, some of which are used to extract energy from the surrounding environment. Life is able to store information, reproduce, and pass on its information to future generations. We could add that life is assembled with a high degree of order from a small number of building blocks assembled using the genetic code of our cells. This code directs the cell to grow into a human ear, or an ear of corn. In this way life resembles the spoken words of a Shakespearean play, brought into existence via arrays of 26 letters that may be meaningless in themselves, but once assembled into words convey an almost infinite variety of information.

Yet, when we look down a microscope at a unicellular organism, darting with its propeller-like flagella across its own tiny ocean held within a petri dish, these simple explanations of life fade away. How can one comprehend the complexity within such a tiny organism? In one small cell there is the control centre of the nucleus, and the organelles for processing proteins, making energy, transporting nutrients around the body, and expelling excess water. Moreover, such cells can assemble into multicellular colonies, with each cell then being charged with a particular task. This seems like a jigsaw puzzle with 4 billion pieces—and with no picture on the front of the box to help get it started. It has, mind you, taken nature around 4 billion years to assemble all of the pieces.

There is, alas, no fossil record on Earth that shows how this complexity began. Almost certainly it involved a whole series of precursor steps, rather than Hoyle's notion of assembly in one single action. The steps may have involved processes such as autocatalysis, in which some chemical reactions become self-sustaining, thus involving a simple form of reproduction that fulfils one of the characteristics of life.[73] For such autocatalytic reactions to ultimately develop into life, though, there is a need to invoke Darwinian evolution so that the reactions evolve to a level of complexity that produces life.

A stumbling block here may be the means of accurately transmitting information from one generation to the next without a genetic code.[74] Without genes there is no easy way to convey good designs to the next generation, and no way to prevent bad designs propagating. To transmit encoded information to your children, a complex molecule such as ribonucleic acid (RNA) or deoxyribonucleic acid (DNA) is needed. But these molecules are complex, and it is almost as unbelievable as Hoyle's aeroplane analogy to believe that they came into existence in a single step, from an array of nucleic acids synthesized in some primordial soup. An alternative hypothesis, therefore, involves the development of a self-replicating proto-RNA molecule. This may have been composed of different 'bases' from the four we all learned in biology at school—adenine, guanine, cytosine, and uracil.

Perhaps its most important quality was an easier mechanism of assembly that was not possible in such complex structures as RNA. Once formed, proto-RNA could undergo Darwinian evolution by coding information that could be passed on to the next generation. That information was selected for stability and versatility, and could ultimately lead to the RNA that is found in the cells of all living things.[75] Whichever of these routes life took—via proto-RNA or by the autocatalytic route, or through some complex interaction of the two—one thing is certain: these processes must have occurred, and so life was born, in water.

Bags of Water

To a silicon-based life form, all life on planet Earth might look like a bag of water. In the Star Trek episode 'Home Soil', Captain Picard and the crew of the starship *Enterprise* are accused of possessing just such a physique by a crystal-based life form on the planet Velara III. How much water there is in those bags depends on the organism, but in humans about 60 per cent of the body by weight is water. That figure hides greater complexity, because of that 60 per cent some two-thirds is within the body's cells, and about one-third is extracellular in fluids such as blood and cerebrospinal fluid. Even between different types of cells in one body, and between different organisms, the amount of water varies considerably,[76] so that a human red blood cell is about 64 per cent water, whereas there can be much less water in cells of organisms adapted to arid environments. One such desiccation-resistant organism is the colloquially known 'sea monkey', a small shrimp (*Artemia*) that is sold desiccated in kit-form to children, the shrimps rapidly reanimating when placed in water. Nonetheless, all cells—including those of desiccated sea monkeys—need some water.

Water is an essential medium in which the chemical reactions of life take place. Water dissolves oxygen, salts, and sugars that provide the means of energy and the vital materials to survive and replicate, and it is the medium in which DNA, proteins, and other components of the cell operate. It is directly involved in primary energy storage from the Sun, as the process of photosynthesis builds sugars from water and carbon dioxide, and yields these again when the sugars are respired for energy. At some point, certainly as early as 3.8 billion years ago and possibly somewhat earlier, life took off. This is evident in the rock record from black organic-rich shales with iron pyrite (fool's gold) that signal the activity of early sulphur-utilizing microbes. These earliest organisms might have been the Archaea, a group that resemble bacteria in small size and 'simplicity', but which have—possibly—yet more ancient origins.

The Engine of Life Turns Over

We cannot see clearly from the fossil record how life first evolved, but we do know that all cells only grow and replicate if there is a supply of energy. On Earth, life is ultimately sustained by the heat and light of the Sun, and to a much lesser degree—at present—by the energy released by radioactive decay within the Earth's interior, an internal heat that is vented by volcanic activity at the surface. In the very deep past, though, the Earth was hotter, and it is plausible that life originated in waters connected to chemical energy from rocks. This is even more plausible if one thinks about the intense meteoritic bombardment of the early Earth, which continued into the Late Heavy Bombardment, and which could have resulted in either the complete annihilation of early surface oceans, or at the very least (for projectiles 'only' 300 kilometres in diameter) left behind only a boiling watery residue.[77]

Amid such impacts, early life might have clung on below the surface, in pockets of water between the rocks more than a kilometre below the surface. Analysis of the time it took for the Earth to cool after the catastrophic impact of Theia (see Chapter 2), and an estimate of geothermal gradients in early continental and oceanic crust, suggest that such habitats existed on Earth as early as 4.4 billion years ago, less than 200 million years after the formation of the Earth. The first surface-dwelling organisms may, therefore, have had ancestors in the underworld: chemoautotrophs subsisting on chemical energy rather than photoautotrophs using the Sun's energy. Chemical signals from the isotopic ratios of oxygen and silicon preserved in ancient marine silica deposits suggest that these subterranean microorganisms would have emerged on to a surface where temperatures remained high compared to the present, and possibly as warm as 70 degrees Celsius. This notion of a warm early Archaean is also supported by evidence from the 'tree of life', in which more ancient groups of organisms have proteins that are stable at higher temperatures, consistent with a thermophilic 'heat-loving' setting for their origins. Life at the surface could only really get going after the Late Heavy Bombardment was complete, and it may have had to survive intermittent periods of darkness caused by the continuing impacts of large meteorites. Earth's atmosphere at this time contained no oxygen, and there was no dissolved oxygen in the oceans either (Fig. 11). Life then had to contend with higher surface temperatures than today, and used metabolic pathways that did not involve free oxygen.

Life in a Noxious World

We can glimpse the ancient environments of Earth through the lives of microbes whose ancestry stretches back some 4 billion years. The Archaea include forms with an ability to tolerate temperatures as warm as 120 degrees Celsius, acidity as low as pH 2, waters that are supersaturated with salt, and the complete lack of oxygen. They evolved on a warm Earth, where the surface temperatures may have reached 70 degrees Celsius and the atmosphere was made of methane,

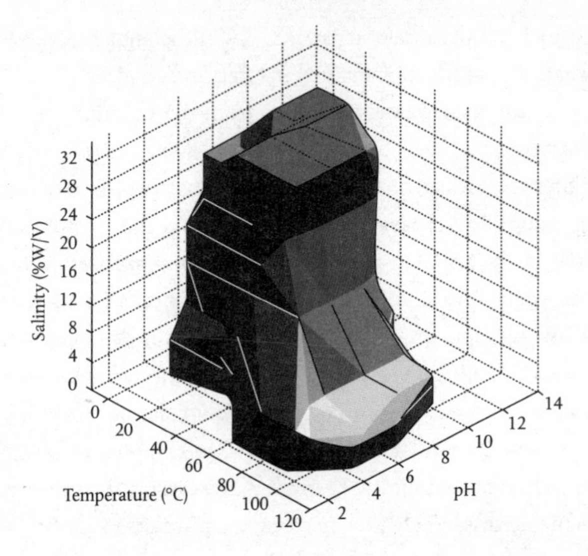

FIG. 11. Not too hot, not too cold, not too salty, not too acid. The envelope of life on Earth constrained by temperature, pH, and salinity. The figure shows the physical parameters of life on Earth through geological time and including the present. Life can exist in environments that are hostile for most living organisms today (for example, hot springs or acid mine waters). However, early life may have had to contend with such regimes across the whole of Earth.

water vapour, carbon dioxide, and nitrogen. They lived in a world where the oceans were saturated in ferrous iron but bereft of oxygen. In such settings, organisms used a variety of means to generate their energy.

Many Archaea are chemoautotrophic, and use energy pathways that are based on inorganic compounds such as ferrous iron, hydrogen sulphide, sulphur, and hydrogen to fix carbon dioxide. Ferrous iron in particular was a ready supply of energy for microbes emerging into the ancient oceans. In modern oceans, ferrous iron is scavenged from water by free oxygen to form insoluble ferric iron (rust), and this gets deposited in sedimentary rocks. But in the anoxic oceans of the Archaean, ferrous iron remained soluble and thus in ready supply, with a continuous source weathered from the volcanic crust of the early Earth.

Perhaps the surface of the early Earth was once pink-red from salt-loving halophile Archaea.[78] There are also the thermophiles, found growing in the scalding-hot water emerging from deep-sea hydrothermal vents. Or the methanogens that colonize regions with little or no oxygen, and which today can be found in rice paddies or the guts of cows and humans. Surprisingly, for the Archaea were once thought to be specialists in colonizing oddball environments, they also live side-by-side with bacteria and eukaryotes in oxygenated waters with normal salinity. They are also widespread as microplankton in the oceans, where they may act as a primary food source. And, for a group of organisms supposedly specialized for heat tolerance, they are abundant in the cold waters of the Antarctic. Nevertheless, despite their modern environmental range, the genetic story of these 'cold-adapted' Archaea suggests their origins lie in anoxic high-temperature habitats, the same habitats that characterized early Earth.

Archaea are also phototrophic, using the energy of light to produce chemical energy that drives metabolic processes in the cell. But this process is not the photosynthesis used by cyanobacteria and higher plants, and it does not liberate free oxygen. Ancient Archaea may also have been heterotrophs, feeding on the inorganic carbon compounds supplied by asteroid impacts; that is a narrow niche, though, with a limited food supply even on the asteroid-bombarded early Earth.

Few places on Earth now resemble the ancient Archaean seas, but Lake Matano on the Indonesian island of Sulawesi may distantly echo those conditions. Matano is a deep lake, extending down 590 metres. It is perhaps 4 million years old, and below its oxygenated surface waters that teem with unique and exotic fish, the lake has no free

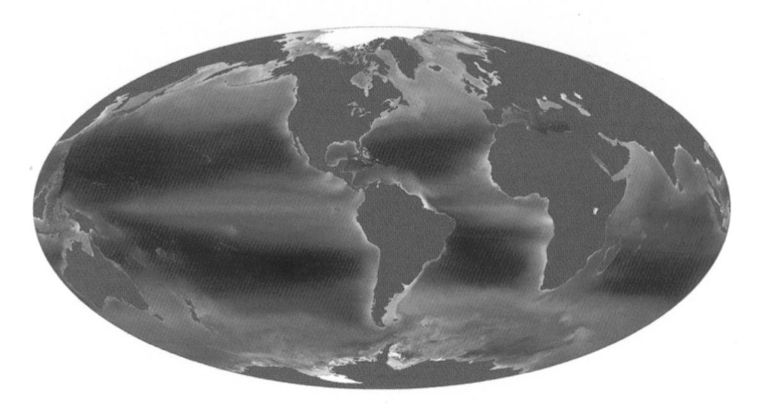

PLATE 1. Chlorophyll concentrations in the oceans: a signature of the phytoplankton.

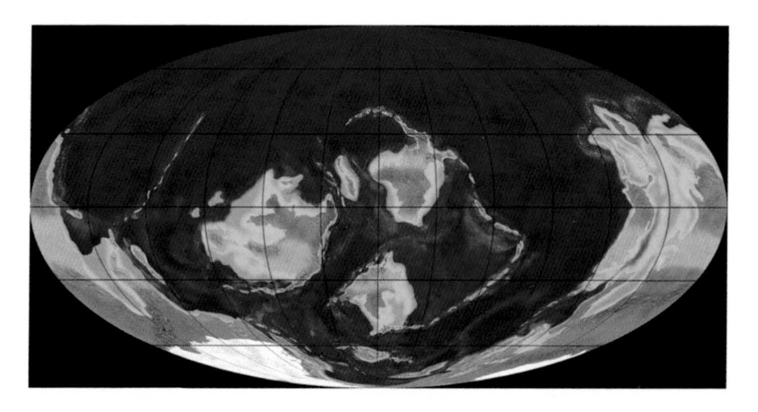

PLATE 2. Oceans of the Ordovician world. In the south lies the giant continent of Gondwana with its polar ice cap. To the north are the ancient continents of Laurentia, Siberia, and Baltica. The northern hemisphere is dominated by a gigantic seaway.

PLATE 3. The fossil ostracod *Nymphatelina gravida* from the Silurian Herefordshire Lagerstätte. The tiny animal, a few millimetres long, is preserved with all of its soft anatomy and, crucially, its brooding eggs for the next generation. The careful parenting that ostracods exhibit may have contributed to their evolutionary success across 500 million years.

PLATE 4. Fish (*Chromis*) and coral (*Stylophora pistillata*) photographed on the healthy Tijou Reef, northern Great Barrier Reef.

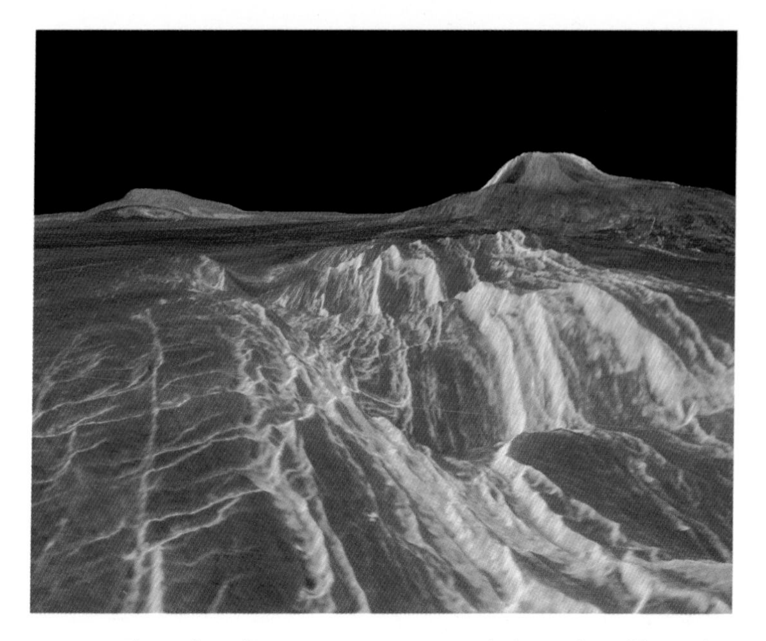

PLATE 5. The surface of Venus as a comparison with the Earth in old age. The view shows a portion of western Eistla Regio. The viewpoint is at an elevation of 1.2 kilometres (0.75 miles) at a location 700 kilometres (435 miles) south-east of Gula Mons, the volcano on the right horizon.

PLATE 6. Mars with oceans: perhaps a view of this planet some 3.5 billion years ago, or in the distant future as the sun grows warmer.

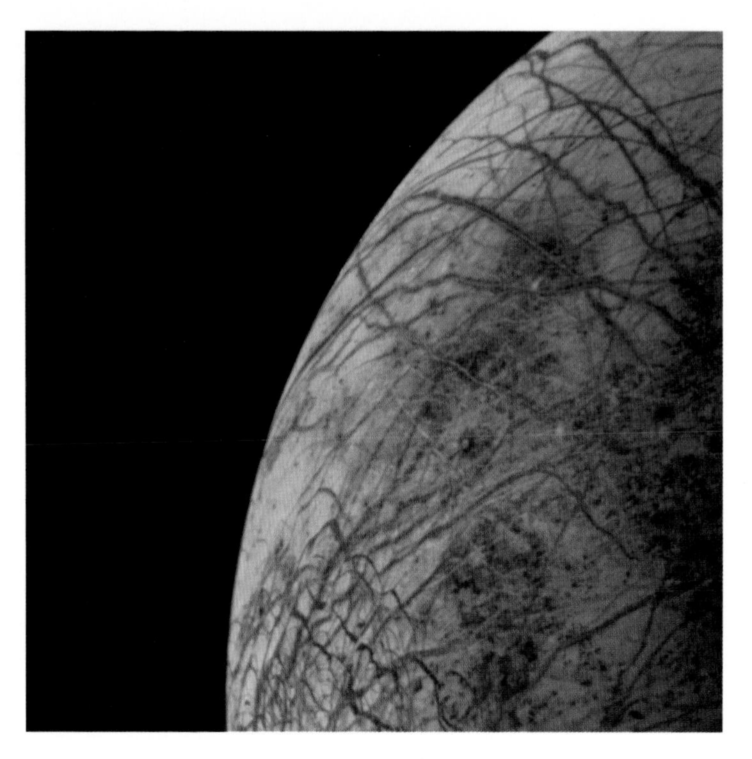

PLATE 7. Surface cracks in the icy carapace of Jupiter's moon Europa, imaged by *Voyager 2* during 1979. The complex streaks on Europa's surface indicate that the crust has been fractured and filled by materials from the interior. The relative absence of features and low topography suggests the crust is young and warm a few kilometres below the surface.

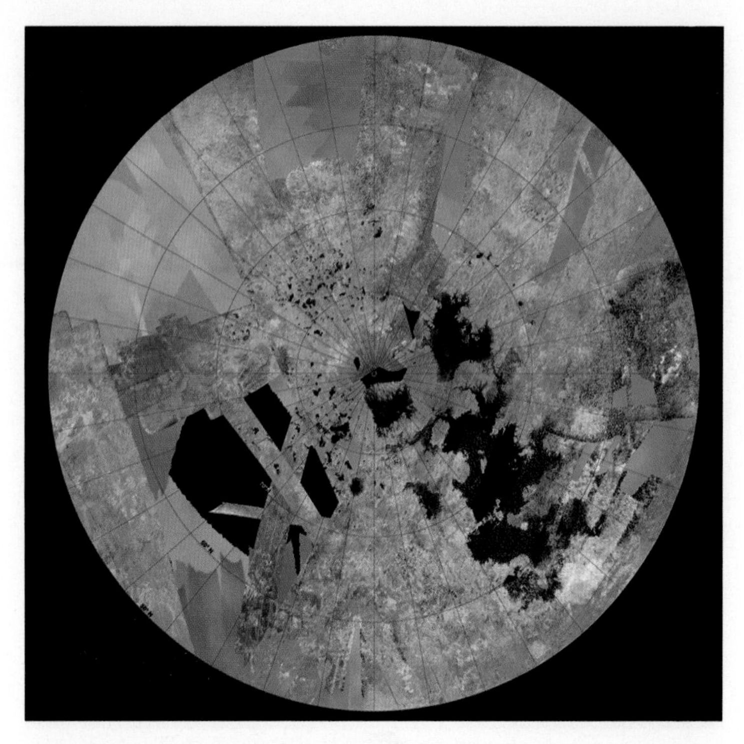

PLATE 8. The northern land of lakes and seas on Titan, the image formed from a mosaic of pictures from NASA's *Cassini* mission. The liquid in Titan's lakes and seas is mostly methane and ethane. The dark irregular form of Kraken Mare, Titan's largest sea, sprawls from just below and to the right of the north pole down to the bottom right. The regular dark shapes are unmapped areas.

PLATE 9. An artist's impression of exoplanet Kepler-22b, which lies some 600 light years away from Earth in the direction of the constellation of Cygnus. Kepler-22b is in the Goldilocks zone of its star, Kepler-22.

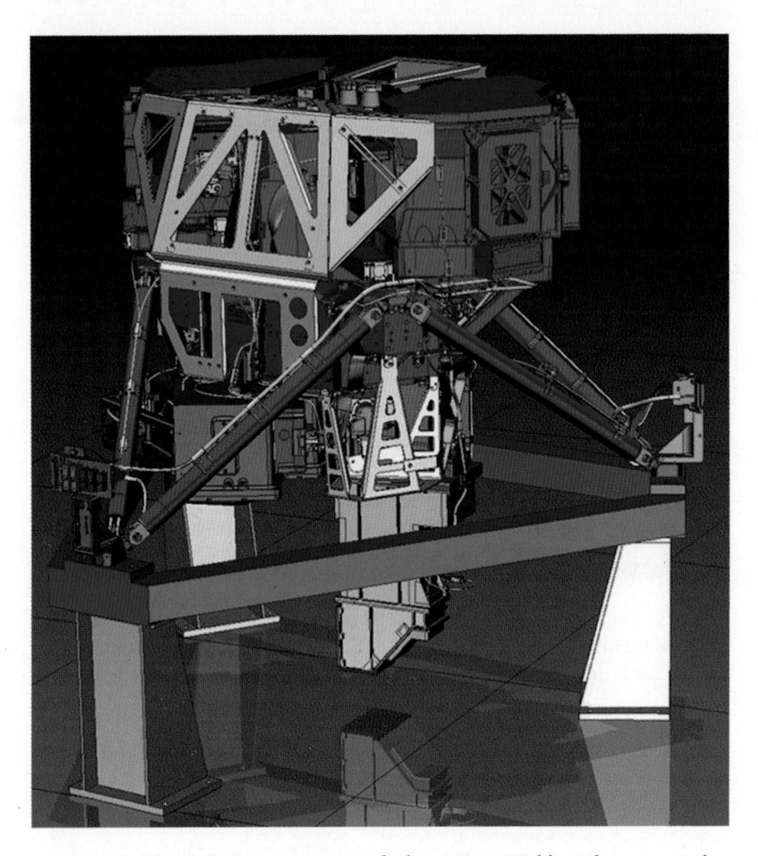

PLATE 10. The MIRI component of the James Webb Telescope under construction: a piece of kit designed to look at the Kuiper Belt, and to look for exoplanets.

oxygen and is rich in ferrous iron sourced from the weathering of lateritic soils in its immediate hinterland. Below the oxygen-rich surface waters of Lake Matano live phototrophic bacteria using ferrous iron for energy, mimicking their ancient ancestors from long ago.

One group of Earth's early bacterial organisms, the cyanobacteria, started a revolution. They began to use water for photosynthesis, combining this with carbon dioxide to make sugars for energy, and as a by-product of the process releasing free oxygen. When this process first happened remains contentious. There is some fossil evidence for cyanobacteria as early as 3.5 billion years ago. However, analysis of the rate at which genetic change has occurred in cyanobacteria, using their internal 'molecular clock', and a wider analysis of the fossil record[79] suggests the rise of cyanobacteria occurred around the boundary between the Archaean and Proterozoic eons of the Precambrian, at a time referred to by geologists as the Great Oxygenation Event, or GOE. This is the point of time, 2.4 billion years ago, when oxygen began to accumulate in the atmosphere and surface of the oceans. It was a toxic poisonous gas for the organisms of the Archaean, but it would have large implications for the evolution of a new ocean state that heralded the Proterozoic Eon.

Stinking Seas

The Earth's oceans have evolved through at least three major states. Most importantly for this story, transitions between these states occurred as a direct response to their interaction with life.

The banded iron formations (BIFs) of the Archaean represent the first ocean state, of ancient anoxic seas full of ferrous iron weathered from the Earth's volcanic crust. Gradually this iron was removed from the oceans by the actions of anoxygenic phototrophs (microbes that photosynthesized but did not release free oxygen), or possibly by some early cyanobacteria. This iron was deposited on the seabed in a series of layers, each separated by a thin layer of mud or chert, so that BIFs, when fossilized in stone, resemble an ancient supermarket bar code. Then, near the beginning of the Proterozoic Eon, a new biological revolution in photosynthesis produced, for the first time, free oxygen. This event is associated with the first great snowball Earth glaciation of the Precambrian, called the Makganyene after the area in South Africa where its ancient sedimentary deposits are preserved. Perhaps Makganyene was a direct response to the oxidation of the greenhouse gas methane in Earth's early atmosphere, plunging the planet into an icy global climate. As oxygen began to accumulate in the atmosphere it also changed the weathering of minerals on land, leading to a host of new mineral forms. Notably, it oxidized iron pyrite to sulphate, which was delivered to the oceans by rivers (see Chapter 4). This sulphate leaves the oceans either as pyrite buried in sediments—the end result of a bacterial transformation deep in the ocean—or as the evaporitic mineral gypsum, calcium sulphate. The influx of sulphate to the early Proterozoic oceans is important, because while the near-surface waters of these oceans may have been oxygenated, the deep oceans remained oxygen-depleted. This was the realm of oxygen-loathing bacteria that converted the sulphate back to its reduced form as hydrogen sulphide: in turn, hydrogen sulphide lowers the solubility of key trace metals such as molybdenum and copper that are essential to nitrogen fixation by cyanobacteria. Sulphidic ocean waters are thus bad for life in general—and not just for oxygen-respiring forms.

In such oxygen-depleted oceans, organisms using anoxygenic photosynthesis in the deep part of the photic zone (the regions to which light reaches) may have used up much of the nutrient supply, thus limiting the biological productivity of photosynthesizing cyanobacteria in the surface waters. It seems to have been thin pickings all round. Ocean life evolved only slowly during the almost unimaginably long interval of time of the early Proterozoic (the 'boring billion' some geologists have termed it), perhaps due to chronic nutrient starvation in the sulphidic seas. Then, something striking happened. About 1.7 billion years ago the first eukaryotic organisms appear in the fossil record. Life had engineered the nucleus to a cell.

The Origins of Marine Diversity

Lynn Margulis (1938–2011) was one of those larger-than-life characters who tended to shake up any scientific meeting she attended, and the panache with which she described the sheer *variousness* of the microscopic life she studied was quite memorable. Defying categorization herself, she noted that life did too, especially when it came to the question of just what is an individual organism.

Bacteria—she said in effect—really could have their cake and eat it, although in reality the process happens the other way around. Endosymbiosis is the posh word for it. A large cell engulfs a smaller cell—or, perhaps the smaller cell is trying to parasitize the larger. Whatever the exact process, the outcome is not always the digestion and dissolution of one by the other. The two original organisms, one inside the other, both survive and begin to collaborate, sharing out the tasks of this new greater whole between them. This is not murder, but cooperation—and it opened up the world to entirely new possibilities.

Lynn Margulis, as a young scientist, had great problems getting her idea published. It was not quite new—it had been mooted as a theoretical possibility for the best part of a century, while she based her thesis on observations of living organisms. Nonetheless, for the day it was a radical, even outrageous, idea. She submitted her paper successively to 14 journals—and received a rejection each time. The fifteenth journal accepted it (Margulis did not lack for tenacity). Today, this scientific battle has been won, and it is recognized as one of the great revolutionary changes in biology. The nucleus in a cell, the energy-providing mitochondria, and the chloroplasts—all are derived from originally independent organisms that went into partnership. Between them and their hosts they produced a new cellular machine: a eukaryotic organism.

For geologists this can be a little depressing, because there is little likelihood of finding fossil evidence of quite *how* this important event happened, although some molecular 'fossil' evidence remains in the genes of modern microbes. Whatever the unobserved, original mechanisms of this process, it clearly led to the ability to construct cells and ultimately bodies with a much wider array of forms and functions. Evidence of *when* this step took place can be found, though, within preserved strata laid down on the floor of the Proterozoic oceans.

Prokaryotic cells such as bacteria can be large. They can have processes that stick out from the cell, and they can have cell structures that preserve as fossils. But no single prokaryotic cell possesses all three of these characteristics, and neither do they possess the complex surface architecture of eukaryotes.[80] For such complexity to develop requires the invention of 'scaffolding': the development of a cytoskeleton. It also requires subdivision by membranes within the cell ('endomembranes'). Both of these structures are characteristic of eukaryotic cells. Based on these pragmatic criteria, the first appearance of eukaryotes is seen in fossils from rocks in China and Australia from about 1.7 billion years ago.

In outward appearance, these early eukaryotic organisms are not terribly spectacular. The earliest of them are considered to be acritarchs, a group long considered to be the resting (spore) stages of phytoplankton, although acritarch fossils likely cover a wider range of biological affinities. In the Mallapunyah Formation of northern Australia there are fossils of *Valeria* with its distinctive 'corduroy' pattern of parallel lines, and in the Changcheng

Group of China one can find *Leiosphaeridia*, a rather dull-looking spherical fossil, but one with a complex double structure to its organic wall. Although unspectacular as fossils, the structures they produced indicate the presence of the subcellular building blocks needed to make, ultimately, a much wider array of shapes and structures. Despite their greater cellular complexity, evolution proceeded slowly without the possibility of complex multicellular structures. Then came the invention of sex.

Bangiomorpha doesn't look like much. Indeed, you might easily overlook this fossil organism as some tiny obscure strands of something or other in the rock. But if you have the precise eye, acute brain, and gently eccentric manner of Nick Butterfield, a geologist at the University of Cambridge, then *Bangiomorpha* is a very significant organism indeed.[81] For a start, it is very old, occurring in rocks of the Hunting Formation in Arctic Canada that are perhaps as old as 1.2 billion years: these rocks appear to be have been deposited in very shallow water on the margins of a sea. *Bangiomorpha* shows different types of cell (Fig. 12). It has structures designed to make a multicellular holdfast for attachment to a firm substrate. From this holdfast arise unbranched filaments composed of multiple cells, and the layered arrangement of the cells in these filaments is directly comparable to that of the modern red alga *Bangia*. *Bangiomorpha* also shows the earliest evidence for sex, with at least two distinct spore-producing phases.

The co-occurrence of sexual reproduction on complex multicellular organisms with specialized cell types may not be coincidental: one may have led inevitably to the other. In the late Proterozoic the diversity of eukaryotic organisms, both unicellular and multicellular, increased dramatically. Sex, it seems, provided a rich source of evolutionary novelty that would lead to large, three-dimensional organisms populating the oceans.

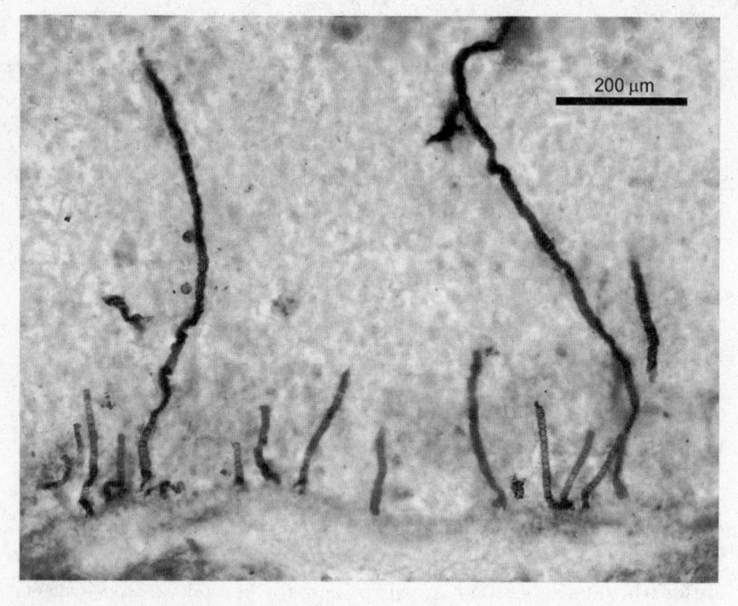

FIG. 12. *Bangiomorpha* is a possible red alga from rocks 1.2 billion years old in Arctic Canada. It shows the earliest fossil evidence for sexual reproduction. The scale bar is 200 microns, one-fifth of a millimetre.

Building the First Bodies

Bodies can do so much more than single cells. They can build three-dimensional structures with tissues, and make shapes that encourage the flow of water (and nutrients) over those tissues. They can make mouths and guts to specialize in feeding, and they can build fins and legs for locomotion to take the organism to its food. Once cells began to cooperate in a single construction, albeit very simply at first, Darwinian evolution could operate to select for structures and specialities that confer some advantage. For multicellular life to progress, though, Darwinian evolution must act to favour some cells becoming differentiated, perhaps to specialize in protection or reproduction. Recent remarkable experiments conducted by William Ratcliff and his colleagues at the University of Minnesota have shown that yeast, the unicellular organism we use to make bread and beer, displays some of the characteristics that may have led to cell differentiation in multicellular organisms. In an elegant and simple experimental design that has echoes of Francesco Redi, Ratcliff and colleagues placed a strong selection pressure on yeast cells to cluster; this was simply gravity, and its effect on the sinking rate of yeast cells or clusters of cells. They saw that cells did not stay isolated, but after only some tens of generations began to stick together to form clusters that sank faster. More importantly, they noted an increase through the generations of the process called 'apoptosis', or 'programmed cell death'. Sinister as it sounds, apoptosis is crucial to multicellular organisms because it allows for the formation of specialized tissues such as fingers and toes. That apoptosis increased through the generations of yeasts indicates the beginning of a division of labour in the colonies—between those cells that are viable for maintaining the colony structure, and those that die to allow the break-away and propagation of new colonies.[82] Such specialization provides a whole new way of adapting to different substrates, water temperatures, salinities, and food supplies. It is also the beginning of a differentiation that would result in the first complex marine ecosystems.

Among the first organisms to show cells taking cooperation further were sponges, which have a good fossil record that extends back to the Cambrian, with sponge-like animals also in the late Proterozoic. They lack specialized tissues for digestion, nerves, or circulation, but they can build shapes that enhance the water flow over their tissues. They also possess cells that can move within the body and change their function.

When did the transition from unicellular organisms to sponges occur? Fossil evidence indicates that sponge-like animals were present in marine ecosystems some 600 million years ago,[83] with perhaps an even earlier origin for animals. This is important, because it suggests that the development of multicellular life in the oceans occurred during the period of Earth's history that is characterized by the intense, world-embracing Late Proterozoic snowball glaciations, when ocean productivity and the nutrient cycle may have been severely curtailed.

Sponges play a wider role. Before they appeared in the oceans there were no suspension-feeding organisms, and so the seas would have been commonly turbid from clouds of suspended living and dead cells and microscopic organic debris. By themselves, such tiny, low-density particles settle to the sea floor only *extremely* slowly, and so can accumulate within the water. Sponges filter-feed particles from water down to cyanobacterial size, and as a result they remove turbidity from the water. How does that help other organisms to thrive? The Proterozoic ocean biosphere largely consisted of cyanobacteria that lived in the surface waters. Because of their small size, and the absence of mechanisms to concentrate them into clumps, much of the organic material in Proterozoic oceans would have remained buoyant and near the surface. This also reduced the penetration of light into the oceans and therefore limited the development of larger eukaryotic and multicellular photosynthesizing organisms.

Sponges not only clean the water, but they accumulate a large quantity of organic matter that other organisms can use for food;[84] in some settings sponges accumulate perhaps as much as half their body mass each day. They can use the dissolved organic matter in seawater to build organic tissues, and they then expel these tissues as filter cells that can be consumed by other reef organisms. Sponges may have been integral to increasing the flow of energy from one trophic level to another, and were therefore an essential prerequisite to building more complex ecosystems. In clearing the water of decaying organic particles they have also been suggested as a means of helping the spread of oxygen through the oceans.[85] One could predict, therefore, that for complex marine ecosystems to develop on other planets some kind of 'sponge filter-feeding organism stage' is essential.

Then, another giant leap. This was to cnidarians, the other marine group considered to have very ancient (Precambrian) origins, and one that includes corals, jellyfish, and sea anemones. The biological innovations here were a simple gut for digestion, and—importantly for interaction with the physical environment—the beginnings of muscle. Cnidarians also possess nerve tissue, and some modern cnidarians have eyes. Crucially, some cnidarians also developed into predators. In time, the act of eating other animals was to produce an intense arms race in the oceans, but in the late Precambrian world this had not yet happened, or at least there is little fossil evidence for it. The fossil record of the Late Precambrian also does not preserve a record of how animals evolved guts and other complex tissues, although molecular clocks suggest that cnidarians were present 600 million years ago.

There are other branches of the early story of multicellular life. Weirdest of all are the 'flat animals', the closest evolutionary affinities of which science cannot agree on. Are they simple animals that are closely related to sponges, or are they the descendants of more complex animals that have lost much of their morphology? Whichever, flat animals, more correctly called Placozoa, are small (millimetre-scale), flat, and irregularly shaped. They resemble amoebae in size and form, although they are multicellular. Imagine a world where animal organization has progressed only to the level of Placozoa. It is a world 'peopled' by organisms quite reminiscent (if tiny versions) of the alien invader of the 1958 comedy-horror movie *The Blob*, not least because placozoans feed by ingesting organic debris as they move along. Perhaps flat animals, and not sponges, represent the beginnings of a multicellular world? With no fossil record, this remains speculation.

The Quilt Creatures

For over 3 billion years, the oceans stirred only to the sound of flatulent microbes releasing bubbles of gas from the seabed as by-products of respiration. Over much of the sea floor there were thin slimy constructions a few centimetres thick, colonized and built by archaea and bacteria. Although thin, these bacterial mats presented a barrier between the water above and the sediment below, and the sediment remained the realm of sulphate-reducing microbes, existing in an environment with no oxygen. Above the mats, sponges had begun to colonize the ocean floor, rising to exploit the supply of organic material in the water just above the seabed. Into this world of microbial mats and sponges came the multicellular Ediacarans, a striking array of new organisms that were a global phenomenon from about 600 million years ago, and which are found in rocks worldwide from Australia to Canada.

The Ediacarans have puzzled scientists since their first discovery, by serendipity, in the county of Leicestershire, England, by James Harley and John Plant in the spring of 1848[86] (forgotten to science, these fossils were then rediscovered 110 years later by schoolboy Roger Mason and Leicester academic Trevor Ford). On a journey to the ruined Leicester Abbey—one of those razed to the ground by Henry VIII's dissolution of the monasteries—Harley and Plant stopped to visit a recently disused quarry in the area now known as Bradgate Park, where they noticed strange circular marks on the Precambrian slates, which they likened in shape to ammonites. The two travellers had discovered fossils in rocks where science then said no fossils could be found.

This is a 'first discovery', because Ediacaran fossils were subsequently and independently 'discovered' in Newfoundland, Canada, by the Scotsman Alexander Murray in the 1860s, and then rediscovered in modern times in the Ediacara Hills of southern Australia by Reginald Sprigg in the 1940s. In each case the fossils were treated with scepticism, coming as they did from rocks that pre-dated the Cambrian System (in which 'normal' fossils such as trilobites first appeared). Indeed, the Australian finds were first described as Cambrian: fellow scientists either would not believe they were fossils, or would not believe they were Precambrian in age. The Leicestershire fossils were simply dismissed, rather snootily, in the venerable *Quarterly Journal of the Geological Society* of 1877 as 'accidental' structures. Yet the 'discs', 'bags', and 'fronds' represent a profound—if ultimately failed—experiment in multicellular oceanic life.

The preservation of the Ediacarans is in itself something of a mystery, for they had no hard skeletons. Perhaps their association with microbial mats is key here, with these preserving an impression of the undersides (and tops) of the fossils as they were buried in sediment. Or, the microbes themselves might have formed mineral templates of the organisms. Even more mysteriously, no one really knows what the Ediacarans *were*, biologically. Many, including the fossil *Ediacaria* itself, are disc-shaped, and early interpretations suggested an affinity with jellyfish, although this now seems unlikely. The famous German palaeontologist Adolf Seilacher likened some of them to 'quilts', biological structures with no modern analogue that may have been able to survive in a world with few predators. The 'quilts', with their broad surface area and lack of internal structures, may have enabled feeding by osmosis from the surrounding water (Fig. 13). Some Ediacarans seem to bear a closer relationship to living animals, and the disc-shaped *Aspidella*, known from Newfoundland and Australia, might have affinities to tube anemones.[87]

FIG. 13. *Charniodiscus* and *Beothukis* on the famous and now carefully protected 'E surface' at the Mistaken Point Nature Reserve, Avalon Peninsula, Newfoundland. The fossils on these cliff tops represent the Ediacaran biota living on a sea floor that was covered in volcanic ash. The erosion of the soft ash above has revealed the impressions of these organisms. The ash has also allowed direct radiometric dating of this rock to 565 million years old. Horizontal field of view is about 56 cm.

How did these early ecosystems differ from the ecosystems that would come to characterize the Phanerozoic Eon—the one that we live in now? As well as the frond-shaped 'quilts' that might have fed directly from the water column by osmosis, other Ediacarans, including the pear drop-shaped *Kimberella,* may have been scratching and grazing on the surface of the microbial mats. Still others, like the elongate *Spriggina,* named for Reginald Sprigg, look superficially a little like arthropods. Many of these organisms were probably making a living feeding on the widespread bacterial biomass. However, none of them penetrated the seabed to search for food, as many modern organisms do. To do that, to be able to burrow, an animal needs a head and tail end, and a bilateral symmetry like a worm, mollusc, or shrimp.

The Ediacarans represent a large step forward in the evolution of ocean life, but they themselves were not to persist, and as far as we know they did not leave descendants. Were they really a failed experiment? Well, they persisted, albeit without showing very much in the way of evolution, for some 60 million years during the late Proterozoic (and perhaps some elements persisted for longer, into the Cambrian). They were then swept away, and the ancient microbial mats disrupted, by a new change in the ocean realm. Organisms with structure and behaviour far more complex than anything yet to appear on the Earth were evolving, and they would invade the world of quilts and microbes.

Tooth and Claw

A walk along the seashore 600 million years ago would be a very different experience from walking along the beach today. A few hours after the tide has fallen you will find no trace of seashells on the shore. No indication of life skipping across the surface of the beach, and no indication of burrowing into the sand below. There is no point in searching for lugworms here, no telltale worm-shaped patterns of sediment being squeezed out like toothpaste above the sand: and indeed, no fish are in the sea to be caught by fishermen. Nothing but a pristine beach of golden sand, with perhaps here and there the signs of microbial mats glistening at lowest tide.

One hundred million years later, and that walk along the beach becomes very different—perhaps not so different from a beachcombing walk of today. Now there are shells everywhere, and some of these even resemble the shells we find on beaches today, such as the burrowing brachiopod *Lingula*, with its two elongated valves and long protruding pedicle. Animals skip across the surface of the sand,[88] and others burrow below its surface. There are few traces of microbial mats now—just golden sand, beautiful ripples, and the traces and shells of animals.

How did life change so fast between 600 and 500 million years ago—for that is a very short time compared to the billions of years of slow change in the Precambrian? We may be tempted to think of Fred Hoyle's analogy again, of a whirlwind ripping through the Earth's oceans to assemble complex life forms from a much simpler set of components. But that misses the point once more, for life had been accumulating a series of steps towards complexity: the development of cells with organelles and nuclei, the acquisition of sex, the organization of many cells into multicellular organisms, and the division of labour between those cells to specialize in reproduction, feeding, and locomotion. Finally, there came the development of the first complex groupings of organisms in the late Proterozoic, remnants of which persist today in the forests of sponges in the deep oceans.

Associated with some of the later Ediacaran assemblages are the first simple trace fossils. They are the first indications of animals attempting to penetrate the seabed to search for food. Among these early trace fossils are those called *Treptichnus*, a three-pronged burrow that shows repeated penetration of the seabed. They first appear in rocks about 550 million years old, but then suddenly became a very important component of seabed deposits 541 million years ago.

These patterns had intrigued Jean Vannier, a scientist at the University of Lyon, France. Perplexed by the complexity of behaviour they represented, and surmising that they might represent the trails of priapulid worms from the abundance of these animals as fossils in certain Cambrian deposits, Jean designed an experiment to test his hypothesis. Priapulid worms form a very small group of animals at present, with only 18 living species. But they are distinctive, in that they have an uncanny resemblance to the human male reproductive organ, and hence their informal name of 'penis worms'. Jean dredged up some penis worms from a Swedish fjord along with some of the mud from the bottom. Back in the laboratory he observed the action of the worms as they moved across this mud. The worms had a very distinctive motion, anchoring themselves by means of small spines in their cuticle and then projecting themselves forward in a series of peristaltic waves. The pattern they produced in the mud was remarkable, in that it closely resembled the patterns of *Treptichnus* in rocks 541 million years old (Fig. 14). Here then, was evidence of worms beginning to burrow into sediment and the opening up of a new world of opportunities for life in the sea.

This explosion of life at the beginning of the Phanerozoic Eon (*our* eon) transformed the oceans. It was the third pivotal biological event, the first being the origin of life itself, the second being photosynthesis, and then this one, often named the Cambrian Agronomic Revolution. As animals began to burrow into the seabed, new sources of food became available, including a vast untapped resource of organic carbon. There were new homes for living in and hiding in, and new patterns of moving and feeding. There are remarkable similarities between this Cambrian Agronomic Revolution and its near namesake, the human agrarian revolution that took place 541 million years later. Both involved new patterns of making food, releasing a surplus of energy that allowed complex new systems to develop. What began with organisms burrowing into the sediment cascaded rapidly into a series of steps towards complexity—and transformation of the ocean environment.

The first Cambrian animals were essentially soft-bodied, and left only the traces of their activity. Then, as competition for resources increased and animal ate animal, mineralized skeletons evolved for protection, and also to support new ways of living. A variety of

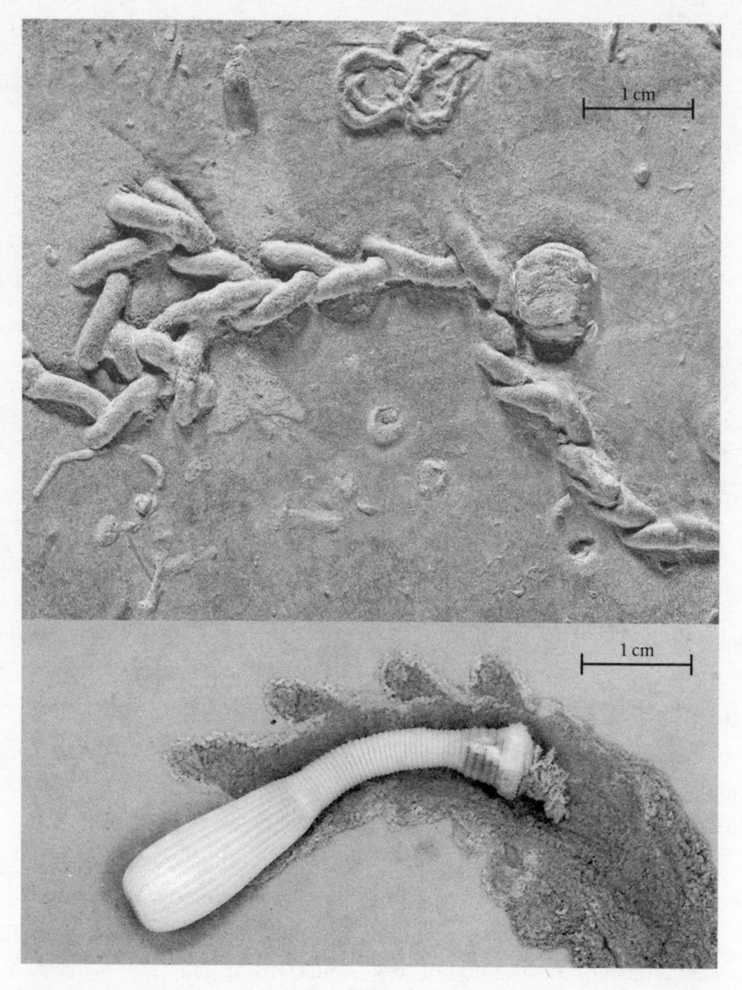

FIG. 14. The modern priapulid worm *Priapulus caudatus* (bottom) making traces like the fossil *Treptichnus*, from early Cambrian rocks of Sweden (top). The scale bars are 1 centimetre.

creatures arose, actively walking or crawling on or burrowing through the sea floor, and helping to mix oxygen into it. Others ascended into the plankton, being swept along by (and helping to stir) the ocean waters. These made bodies and skeletons of carbon, nitrogen, phosphorus, and calcium, and so held a significant part of the ocean chemistry in their bodies. Other creatures, the corals and algae, would eventually build monumental constructions—the reefs—on parts of the sea floor, hotspots of biodiversity that could last for millions of years and lock away billions of tonnes of carbon within them.

Fossils represent a tiny fraction of the organisms that have lived in the Earth's oceans, and of these the vast majority only preserve the hard skeletons. The soft tissues—the eyes, brains, muscles, and internal organs—rot away quickly. Even more recalcitrant tissues, like cartilage, have a relatively short shelf life after death. Only rarely do the soft tissues become preserved, but in an almost perverse but illuminating act of nature many of these very rare fossils occur in rocks of Cambrian age. We probably know more about the soft-bodied animals that lived in the oceans of the 500 million-year-old Cambrian Period than we know about life in the oceans of, say, the much more recent Miocene Epoch, just 25 million years ago. In those ancient Cambrian fossil-bearing rocks, some of those first animals have preserved the soft bodies almost intact. Chief among the sites with soft body preservation is the fossil fauna of Chengjiang, named after the small provincial county in southern Yunnan Province, south China.

Cambrian fossils had been known from Yunnan since the early part of the twentieth century, when the self-trained French palaeontologist Henri Mansuy passed through the region in 1903 to 1904. His survey was not entirely academic, as the French authorities in Hanoi wished to extend their influence into China by building a railway to Kunming. Nevertheless, Mansuy was able to collect a number of fossils, including trilobites, which suggested the rocks here would supply rich pickings for palaeontologists. Some 80 years went by, during which time the political landscape of this region shifted dramatically, and a young Chinese geologist working at the Nanjing Institute of Geology and Palaeontology now passed this way. His name was Hou Xianguang, and he was intrigued by a group of small arthropods called the Bradoriida. Mansuy himself had collected a few specimens of these animals that were once thought to be fossil ostracods (seed shrimps). Although not a native of southern China—he was born in Jiangsu Province in 1949—Hou Xianguang was to make one of the most extraordinary fossil discoveries in history. It was a discovery that would elevate Yunnan Province into the premier league of fossil localities.

It was during the rainy season, on 1 July 1984, that Hou Xianguang left his lodgings at Dapotou village. He had hired a local farmer and cart to help with the excavation and transport of the fossils (Fig. 15).

FIG. 15. Hou Xianguang (wearing the white hat) on the way to collecting Chengjiang fossils in the summer of 1984.

The two took the 5-kilometre walk up to the hill called Maotianshan where they had been collecting. It seemed like an average morning on that quiet road, with the occasional passing farmer. Hou had with him an umbrella and the farmer carried a plastic sheet to keep off the rain, although he recalls that often his legs got very wet below the knees. After a breakfast of noodles, Hou was in good spirits, and carried with him some steamed bread for lunch. As ever, he was searching for the enigmatic little bradoriids, when at about three o'clock in the afternoon a slab of rock broke open to reveal a trilobite-like animal called *Naraoia*. Hou Xianguang had seen trilobites many times before in these rocks, but this one was different: Hou could see immediately that this fossil had legs. He was dazzled by what he saw before him—it seemed as though the animal was alive, as if it was trying to swim across the surface of the mudstone. He carried on working that day until dark, making the walk back to Dapotou village—a hazardous journey in darkness—along the winding mountain road. Hou was not worrying about the dark. That night he could barely sleep for excitement, as he realized that he had discovered the oldest soft-bodied Cambrian fauna in the world, one so old that it had existed only shortly after animals themselves had originated.

Chengjiang is truly remarkable. Here, in rocks over 520 million years old, are the bodies of a multitude of arthropods—from the familiar trilobites, to the much weirder-looking *Fuxianhuia*. Preserved even with its brain, this creature looks a bit like a large sea slater with a long tail and stalked eyes (Fig. 16). Then there are the arthropods such as *Fortiforceps*, with its large stalked eyes and protruding forceps-like frontal pincers—clearly a combination evolved to help hunt and eat other animals. Here too there are worms, including many priapulids; lobopods that look like modern velvet worms; jellyfish-like cnidarians, sponges, and animals that are more familiar to Cambrian rocks everywhere, especially brachiopods (lamp shells).

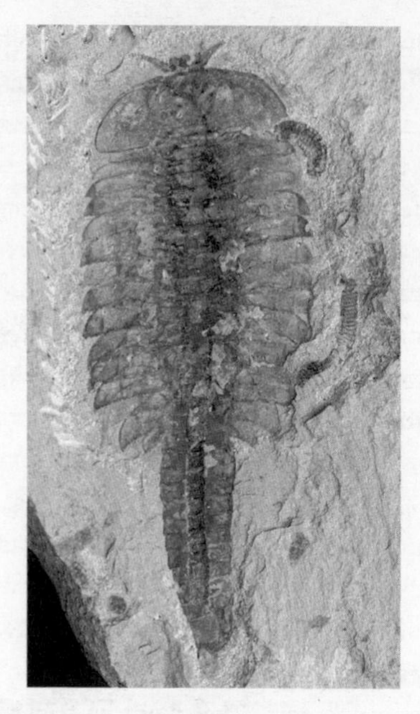

FIG. 16. The exceptionally preserved arthropod *Fuxianhuia* from the early Cambrian rocks of Chengjiang County, Yunnan Province, south China. Up to 11 centimetres long, this animal was probably a predator.

Among the bottom dwellers are predators, feeding on the tiny bradoriids: we know this for certain, because we find fragments of bradoriids preserved in their fossilized guts. A few Chengjiang animals—the jellyfish and some of the arthropods—suggest that the water column was being colonized. This colonization of every level of the water column was to have profound implications for the development of life through the Phanerozoic. Fast food—as we shall see—was on the way.

These early Cambrian ecosystems were not only complex; they varied across the world. In the equally celebrated Cambrian fossil locality of the Burgess Shale in British Columbia, one can find the weirdly shaped (and named) wiwaxiids. Looking like an animal with a penchant for punk rock, its spiky 'hairdo' perfectly matches an organism whose affinities remain enigmatic—it might be an armoured (polychaete) worm, or perhaps a mollusc, or perhaps neither.

Among the Chengjiang fossils there is also a little fish-like animal, called *Haikouichthys*. The first part of its name is from the locality, near the town of Haikou in Yunnan where the fossils were first found. The second part of its name, 'ichthys', means 'fish'. There is a debate among scientists about whether fossils of *Haikouichthys* preserve enough information to say that this is a true fish, but that need not trouble us. For *Haikouichthys* does preserve gills, a notochord, a dorsal fin, and a distinctive head. Small and insignificant in this world of Chengjiang organisms, it nonetheless shares a common ancestry with all of the fish living in the seas today, with the blue whale, and also, of course, with us.

Becoming Complex

If Cambrian marine ecosystems were inherently more complex than those of the Precambrian, they were also unstable. Much of the ocean's life remained in a narrow belt on the continental shelves fringing the ancient continents: the deep oceans remained oxygen-poor and probably sustained few large multicellular organisms. Cambrian ecosystems were also species-poor by comparison with those of modern seas. This may have contributed greatly to their vulnerability. Conservation biologists often talk of 'keystone species', a term coined by Washington University zoologist Robert Paine in a now classic paper published in 1969. The effect of keystone species on the ecosystems in which they reside is profound. In marine settings, keystone species range from top predators, such as sharks, to more humble creatures such as sea urchins and starfish. The purple sea star is a classic keystone animal. It lives in the shallow tidal zone off the coast of northwest America. It is not the top predator in its ecosystem, and indeed it is relished by sea otters. It in turn consumes the mussels of the intertidal zone. The presence of just a few sea stars can keep mussel populations down to levels in which they form only part of a diverse intertidal community. Remove the sea stars, and the mussels rapidly multiply to become dominant, crowding other species out.

In ecosystems where there are many species, vulnerability to the loss of a few keystone species tends to be less. In Cambrian ecosystems, with perhaps a few tens of species, disruption of one or more of the keystone species may have collapsed entire ecologies. Examination of the fossil record of the Cambrian shows a number of major extinction events, perhaps as many as six within its roughly 50 million year duration, and many more than at any other comparable time in the Phanerozoic Eon. These extinctions affected familiar groups such as trilobites, and less familiar groups such as the reef-building but sponge-like archaeocyathids, which flourished briefly in the tropical oceans worldwide, before most went extinct before 510 million years ago. Environmental factors may have played a major role in these extinctions, as the anoxic waters of the deep oceans periodically spilled on to the continental shelf, suffocating much of the life there.

It is a signal of the increasing sophistication of life in the Cambrian seas that such mass suffocation events may have spurred the rise of other groups of organisms. The biological pump is the engine that delivers carbon from the surface waters to the deep oceans, and it is thus an integral part of the ocean food supply, and of the carbon cycle. Seabed anoxia, in the Cambrian and elsewhere, is implicated in forcing many organisms into the water column, especially those that had the right pre-adaptations, such as a life cycle with planktonic larvae, strong swimming abilities, or a penchant for buoyancy. Once established, this zooplankton began to release a stream of faecal pellets from the surface to the depths, together with discarded gelatinous feeding nets that some of them produced and—ultimately—their own bodies. To us as fussy humans it is not the most appetizing food supply (oceanographers call it 'marine snow', or more bluntly the 'faecal express'), but it is one of the main reasons why the deep ocean is not 'azoic'—or lifeless—as that amiable Victorian-era rebel Edward Forbes thought (see Chapter 5). Indeed, this constant stream of poo maintains the ecosystem of the deep ocean. Without it, much of the organic material fixed by phytoplankton would stay in the surface layers and never reach the seabed. Only when it is clumped together does it lose its buoyancy and sink, and it is the role of zooplankton and of its predators to deliver this vital resource to the sea bottom.

The Cambrian closed and the Ordovician began. A rapidly evolving and diversifying plankton in the water column may have seeded yet greater complexity on the seabed. In these Ordovician ecosystems there lived myriad marine phytoplankton providing the primary food, zooplankton eating them and being eaten by arthropods, that were in turn preyed on by larger arthropods and nautiloid molluscs, all contributing to a faecal stream that fed the organisms at the seabed. It was, therefore, a seabed that crawled with life. Snails fed on the detritus falling from above, while predatory sea urchins and starfish prowled around, and brachiopods, bryozoans, and corals attached themselves to any hard surface available. The sediment below stirred too, to the movement of burrowing worms and arthropods, and myriad tiny organisms lived as meiofauna, hiding between the sediment grains. Above the seabed, forests of sea lilies swayed elegantly in the current, filtering edible particles from the water. This is a marine world in many ways like our own, peopled with hundreds of species. It is the time that geologists call the Great Ordovician Biodiversification Event (or GOBE for short), when worldwide a much greater abundance of shelly fossils appeared in the rock record. It produced ecosystems that were resilient: ecosystems that could only collapse as a result of events of singular ferocity, such as massive climate change, giant asteroid impacts, and perhaps, much later, the evolution of large-brained apes.

Resilient Life

Once the ocean ecosystem had formed its complex economy, cross-linking those species that produced and those that consumed, it became resilient. These ecosystems now possessed hundreds or thousands of species, so the loss of a few by extinction mattered little. The fossil record usually pays scant justice to the burgeoning biological extravagance of these oceans, showing only the skeletal remains of the head of a trilobite here, or scattered brachiopod shells there. Just occasionally, more remarkable fossil discoveries allow us to dive beneath the waves of these ancient seas, to witness first-hand the remarkable complexity of these newly emerged ecosystems.

One such fossil locality lies near to the small Herefordshire town of Kington, close to the border with Wales. It is called the 'Herefordshire Lagerstätte'. Lagerstätte is a German word that translates as 'a place of storage'. In palaeontology, it means a site with extraordinary fossils. The Herefordshire Lagerstätte is, indeed, a storehouse for some of the most remarkable fossils ever discovered. They date from the Silurian Period, just after the phenomenon that was GOBE. It represents an ocean biosphere in full bloom.

It was Bob King, self-trained mineralogist and erstwhile curator of geology at Leicester University, who first picked up the tennis ball-sized nodules from the quarry near Kington. He had been rooting for minerals in the quarry spoil. Bob picked up the nodules and thought them interesting enough to bring back to the Leicester collections. What he originally saw in them no one is quite sure, and they lay gathering dust in a cupboard for years. The next curator of geology, Roy Clements, stumbled across them one afternoon and suspected there might be something more to these nodules than met the eye. He passed them to his colleague David Siveter, a scientist who specializes in the study of a group of small arthropods called ostracods. David immediately looked at the nodules through a microscope. He then looked up at Roy with some excitement. 'It's got legs,' he said.

There turned out to be legs, and heads, and bodies, and tails a-plenty in those nodules, and hairs on the legs too. By painstakingly cutting the nodules into thin slices, recording the pattern seen in each slice, and then restoring the whole animal in three dimensions as a virtual fossil, David Siveter and his colleagues—identical twin brother Derek at Oxford, Derek Briggs at Yale, and Mark Sutton in London—were able to begin the job of documenting the Herefordshire fauna. They went back to the quarry where Bob King had made the first discovery and traced the nodules to a layer of volcanic ash that fell on the seabed some 425 million years ago. It was this ash that had entombed the Herefordshire animals and frozen them in time, like an ancient Pompeii of the Silurian seabed.

Within the Herefordshire deposit there are fossil relatives of modern horseshoe crabs, hairy worms, delicate crustaceans, striking starfish, and sensational sea spiders, all preserved in three dimensions. There are even brachiopods, with their fleshy holdfasts intact, and also those small arthropods called ostracods beloved by David Siveter. These animals have been known as fossils for two centuries, although only from their hard skeleton—the carapace of calcium carbonate that survives after the death of the animal. But in the Herefordshire Lagerstätte these fossils also show exquisitely preserved swimming appendages and—to the delight of the media—their impressive reproductive organs: ostracods may be small, but males possess the largest penis and sperm relative to their size of any animal (the English tabloid newspaper the *Sun* reported it—and we merely pass this on without further comment—as 'The World's Oldest Todger'). Even their parenting skills are preserved, for ostracods brood their young within the carapace, carefully nurturing the next generation: in the Herefordshire nodules both eggs and young are preserved inside the carapace. The ocean world had become, in many respects, surprisingly modern (see Plate 3).

Marine life survived the five mass extinction events of the Phanerozoic, each characterized by a more than 50 per cent loss of marine species. It survived the greatest environmental calamity of all, at the Permian–Triassic boundary 250 million years ago, when more than 95 per cent of marine species went extinct. And yet this extinction, and those that followed, could not extinguish the life of the seas, or its complexity. No phylum has ever gone extinct since the Cambrian, and a few million years after each mass extinction event the complexity of the ocean food web essentially reconfigured itself. It witnessed the rise of new innovations, such as the development of predators with shell-crushing mouths, and of new organisms colonizing the plankton, such as foraminifera; it witnessed the colonization of the deep ocean.

The journey through the Phanerozoic seas witnessed, too, the colonization of land by plants and animals. The greening of the land took place quickly—as with the Cambrian radiation of life, a mere few tens of millions of years made a lot of difference. When the animals preserved in Chengjiang were hunting and being hunted in the Cambrian, the land was still barren. Only when the sophisticated and modern-looking Herefordshire animals inhabited the Silurian sea floor was life creeping on to land, in the form of pioneering green stalks, and a few hardy millipedes. By the middle of the succeeding Devonian Period, true forests had evolved, covered by all manner of scuttling, jumping, and biting things. From then on, the biosphere truly included a land-based extension. This was in its infancy, though, while its oceanic parent was already very old: not that it still didn't have a few tricks up its sleeve—to take back, for instance, part of what it had given.

Life on land originated, thus, from marine antecedents. In time some land animals would return to the sea. They would become the most magnificent organisms that have ever lived.

Magnificent Whales

It was some 16 million years after the calamity of the end-Cretaceous extinction event, and after the demise on land of the dinosaurs, and in the sea of marine reptiles such as the ichthyosaurs and plesiosaurs, and of the beautiful coiled ammonites. The oceans had changed in other ways too. Planktonic micro-organisms with calcium carbonate skeletons had evolved, their tiny skeletons now raining down endlessly on to the sea floor to make up the pale oozes that, over a century ago, turned up in bucket after bucket of HMS *Challenger*'s laborious dredging programme (see Chapter 3).

Along the luxurious tropical coastline of an ancient India lived *Pakicetus*, a four-legged mammal. *Pakicetus* looked rather like a long-legged large rat: one big enough to frighten the living daylights out of your average domestic cat. It stood knee-high to a human, and was a little over a metre long. *Pakicetus* may have lurked in the shallows, hidden in water up to its snout and able to pounce, wolfishly, on both marine and land-based prey. The ancestors of *Pakicetus* were hoofed animals that kept firmly to the land. Its relatives though, were anything but. The telltale structure of the ears of *Pakicetus* are the clue to its family connections. The overall shape of the skull in this animal indicates that it heard sound like other land-based mammals. But *Pakicetus* possessed a very robust tympanic bulla—the bit of bone in its skull used to conduct sound to the ear—and in this case adapted to conduct sound underwater. The robust structure of the tympanic bulla in *Pakicetus* signals a strong relationship with cetaceans, the group that includes dolphins, whales, and porpoises, and *Pakicetus* is therefore seen as an ancestor of modern whales.

From *Pakicetus* to the first baleen whales is a journey through some 25 million years, one that sees the loss of legs and the development of fins and blowholes, and ultimately of large brains and large size. The journey stretches from the muggy tropical world of the Eocene 50 million years ago to the cooler world of the latest Oligocene some 25 million years ago, and then on to the present. Baleen whales are descended from toothed ancestors. Their rise, to be the largest creatures that have ever lived, is inextricably linked with another great environmental change on Earth: the rise of the bipolar glaciation that characterizes our world now. Unlike their toothed ancestors, baleen whales are filter-feeders, using their comb-like baleen—the material once used to make Victorian corsets—to trap zooplankton, taking in many thousands at a single gulp.

The world of the late Oligocene, 25 million years ago, was already cooling. The first Antarctic ice sheets had formed some 8 million years earlier at the boundary between the Eocene and Oligocene epochs, likely due to global changes in the level of carbon dioxide in the atmosphere, but also influenced by the increasing geographical isolation of Antarctica as Australia moved away to the north on its plate tectonic journey. As South America also pulled away, this allowed the Antarctic Circumpolar Current (the ACC) to develop and strengthen. This is a fearsome ocean current, and one that is perhaps known better by its sailor's term the 'screaming sixties'. Unimpeded by any continent or landmass in its way, the ACC tears around the Earth at latitude 60 degrees south, clockwise from west to east. Sailing across it is an experience of mighty waves and rolling hulls—and serious seasickness, as many Antarctic-bound scientists have experienced. Surface temperatures fall to zero degrees Celsius, and fall farther still as the ocean surface freezes. Here, the Earth's ocean currents that we visited in Chapter 5 are implicated in forcing an impressive change in the Earth's biosphere.

Cooling in the Southern Ocean 33 million years ago had a dramatic effect on the ecosystem. Evidence from fossilized skeletons of tiny phytoplankton called dinoflagellates shows a change to forms that could exist in conditions with winter sea ice: the sea freezes at minus 1.8 degrees Celsius, giving a strongly seasonal signal of plankton blooms each time the ice melts. The seasonal change caused a bloom each year in the surface ocean primarily produced by phytoplankton. The bloom cascaded through the food chain, with zooplankton such as krill eating the phytoplankton and thriving sufficiently to feed whales in turn.[89] As this new ecology developed, the balance between the whales at the top of the food chain and the primary producers at the base of the food chain became intertwined in an echo of the complexity that developed in the Cambrian ecosystem. For whale poo is rich in iron, a most valuable nutrient for phytoplankton in waters that are normally iron deficient. Whale poo therefore gives back to the ecosystem what it takes out, helping to sustain the Southern Ocean ecology.[90]

In size, the largest whales, such as the majestic blue whale, dwarf all other creatures both on sea and on land. From the first Archaea, less than five one-thousands of a millimetre in diameter, to an animal 30 metres long and weighing as much as 170 tonnes, there is an unbroken line of 4 billion years based on a continuous, life-sustaining ocean.

Earth's Oceans: A Template for Other Worlds?

It took 4 billion years of biological evolution on planet Earth to produce the blue whale. The galaxy in which we live, the Milky Way, is estimated to be 13.6 billion years old, only a little younger than the estimated age of the universe at 13.8 billion years. In that time, many terrestrial planets with water will have emerged, and some of these fall in the 'Goldilocks' zones of those stars they orbit. It seems that, with the right chemistry, life might arise on these planets quite quickly from some precursor proto-self-organizing substance. If Earth can be used as a guide, once life gets hold it can spread rapidly across the surface of such planets.

There may be, and may have been, many such worlds with life in the universe. Most of these will be bacterial, and most will need to possess large, long-lasting bodies of liquid water that we may call oceans. Then—even with billions of years of time and relative environmental stability—the leap of complexity to form eukaryotic organisms may never occur, for we don't know how it occurred on Earth. A few worlds may have made this giant leap, invented sex (or some equivalent), and produced planets with sponge-like organisms to filter-clean their seas as a precedent to the evolution of more complex life. Here and there, and likely very rarely, a few planets may have undergone their own 'Cambrian explosions'. Would these rarest of planets have ever produced a Darwin to contemplate the origin of their own kind? We have not yet detected the kind of far-distant radio signals that might suggest such a thing has happened. An answer can only be found by the discovery of life beyond our planet. And *that* is predicated on the discovery of liquid water. In Chapters 9 and 10 we visit oceans on other planets and explore the possibility for life within them. First, though, we have to see what the future might hold for the oceans of Earth.

7

Oceans in Crisis

The Fish in the Sea

Scientists do not like anecdotal evidence, by and large. Stories, after all, grow larger in the telling and retelling. But now and again there is serious purpose in poring over stories from olden days: as, in this case, told by explorers and seafarers. Humans have been fishing the seas for far longer than they have been recording them scientifically, so to get an idea of what the oceans might have been like before we began to change them the old anecdotes can provide useful clues. The stories of what the seas used to be like, centuries ago, are remarkably consistent. The seas, then, were alive with fish. Even the pirates said so.

Captain Henry Morgan lived in the golden age of piracy, when his ruthless—even by piratical standards—exploits were, for the most part, encouraged by the British government as a means to get the upper hand in their colonial rivalry with the Spanish in the Caribbean. He had a barber-surgeon for a time, Alexandre Exquemelin (variously also known as Esquemeling and Oexmelin), who became something of a confidant—and who had literary ambitions. In 1678, Exquemelin published a book, *De Americaensche Zee-Roovers*, a kind of biography of the piracy he had witnessed. It outraged Morgan (who sued over it), but amid the tales of battle and plunder there is also a picture of a virgin sea.

Green sea turtles then, he said, abounded to the extent that ships, when losing their bearings, could navigate by the sounds they made as they swam in countless numbers to the islands to lay their eggs. Today, there are few green turtles. But can a pirate be a reliable source? Take Charles Darwin, then, a century and a half later. Arriving at the Galapagos islands, he noted that the bay where he landed was 'swarming' with fish, shark, and turtles 'popping their heads up in all parts'. Among the fish he noted as particularly common was the local species of grouper. It is now on the Red List of endangered species.

The sea used to be thought of as an endless, inexhaustible resource. No more. They are now, as regards fish, unrecognizable from their original pristine state. In fact, their pristine state is now something of a mystery, because overfishing before the days of scientific records had already changed the baseline so much that we do not know what pristine is. The fisheries scientist Daniel Pauly has called this the 'shifting baseline' syndrome. Hence, the value of the old anecdotes in giving us an idea, however imprecisely, of what the oceans used to be like.[91]

By any measure, the story of decline is all too clear. It is the top of the oceanic food chain that has seen the most change. Take sharks, for instance. There are few creatures that induce more visceral dread in humans. Yet sharks have been a key part of the marine food chain for more than 400 million years. Their fall has been dramatic. In the north-west Atlantic, for instance, declines in almost all shark species exceeded 50 per cent (while some were over 75 per cent)—and that is in just 15 years or less at the end of the twentieth century, since when the plunder has continued largely unabated.[92] This follows the already substantial declines from previous fishing (that shifting baseline again), and so the real declines exceed 90 per cent,

and often are closer to 99 per cent. A number of species, in
iconic forms such as the hammerhead, are at real risk of extinction.
Many of the sharks have been taken just for their fins, for the famous soup (for which the fins only add a gelatinous texture, while the taste comes from nothing more exotic than chicken broth). It is extraordinary that a culinary quirk can have such an effect on the biology of the Earth.

A little farther down the food chain, historical accounts painted a picture of astonishing abundance of the Atlantic cod off Newfoundland. Fish catches, high throughout the twentieth century, peaked in a burst of high-intensity fishing in the 1970s, then crashed in the 1990s as the fish effectively disappeared, collapsing to less than 1 per cent of its (pre-baseline) population. It still has not recovered, not least because of a predator–prey reversal: the now-scarce young cod are being eaten by the forage fishes and invertebrates that used to be hunted by the abundant mature cod of past times.[93] Remarkably, Jules Verne predicted the Newfoundland cod collapse in his classic underwater adventure *20,000 Leagues Under the Sea*.[94]

Living in deep waters these days affords little protection. Five species of commercially exploited deep-sea grenadiers, skates, and eels were assessed at the end of the twentieth century. In the 17 previous years, population declines ranged from 87 per cent to 98 per cent.[95]

The story across the seas is depressingly consistent. Most fish stocks are being overfished, some to the point of functional extinction. It is understandable: fish is a wonderful source of protein, and there are ever more hungry human mouths to feed (and ever more powerful and technologically enhanced fishing fleets). The ecological structure of the ocean is being altered, too, for one cannot remove the top parts of a food web without the effects rippling through to the levels below.[96]

It is not just the sea, though—the sea floor, too, is being refashioned.

question of casting out a net, or a line with
re a lot of tasty things that live on the sea floor,
ne sea floor sediment, lying in wait for passing prey
n predators. When it comes to finding food, humans
ys been ingenious. As far back as the fourteenth century,
E n fishermen had devised a 'wondyrechaun', a long, heavy iron bar with a net attached that, towed behind their boats, was dragged over the seabed. The net was close-meshed so that no fish in its path 'be it ever so small' could escape. It was crude, but effective. The practice became profitable and popular. It was also damaging, as it wrecked the oyster beds, ploughed up the sea floor, and fouled the water, killing off the spat that mature fish fed on. A petition was put to the King, Edward III, who appointed a commission to look into the problem (the answer, they suggested, was to move the problem into deeper water farther offshore).[97]

That was only the beginning. Things have moved on since then, and took a giant step with the widespread adoption of powered fishing vessels in the mid-twentieth century. Now, in terms of proportion, the continental shelf has been more widely ploughed by the ever-active trawling gear than the land surface has been ploughed by tractor. The process rakes over the seabed, the bulging nets bringing up a mass of everything that lives there. The fishermen pick through the dead and dying organisms, tossing overboard everything that they don't consider valuable (generally most of the catch) and keeping those fish that have market value. The oceanographer Sylvia Earle has likened the practice to bulldozing forests to catch squirrels.

In recent years, the trawlers have moved into ever-deeper waters, to resculpt the sea floor there. For instance, a certain tasty deep-sea shrimp, *Aristeus antennatus*, thrives in the deep waters of the north-west

Mediterranean, down to nearly a kilometre below the sea surface, particularly around the craggy topography of the submarine canyons that snake across the sea floor. Every day, hundreds of motorized fishing boats from the Catalan sea-ports drag weighted trawl nets across those canyon sides, to scoop up whatever lives in those deep-sea sediments. The effect on those canyons, as viewed in sonar images, is little short of astonishing. The rugged topography has been smoothed, as if by a giant hand, completely effacing what had once been a complex system of tributaries leading into the main canyon floor.

It *is* a form of bulldozing. The effect on the sediment is a little like what happens when running water passes over a mixture of sand and mud: the mud is swept away, and the remaining sediment becomes progressively more sandy. Clouds of muddy water are raised in the wake of the dredges, and these travel downslope or downcurrent, to eventually settle in quieter, deeper parts of the sea floor, away from the trawled areas. On steeper slopes, things can get more violent. Slabs of seabed can be dislodged to break up into rapidly moving dense slurries that sweep down into the canyon floors and from there be funnelled into yet deeper waters. It is a kind of grand, large-scale sedimentary sorting process, creating new kinds of sedimentary strata.

On the sea floor, where the trawls have passed over, there is commonly a distinct layer, a metre thick or so, in which the sediment grains get larger nearer the sea surface because the finer particles have been winnowed away. This 'coarsening-upwards trend', as geologists call it, is a direct signature of the ploughing of the sea floor. Downslope, particularly in the floors of submarine canyons below the trawled areas, there may be layers of sand or mud, each usually a few centimetres thick. Each of these is the result of a dense turbulent sediment-laden cloud—a turbidity current—that sped downslope, hugging the sea floor, in the wake of the trawlers. Turbidity currents are nature's chief way of transporting sediment from shallow to deep water, and the resulting deposits ('turbidites') are common among the Earth's strata (on the steep slopes, turbidity currents have carved out the submarine canyon systems too). Hence, as a side-effect of trawling, humans are now mimicking one of nature's patterns. More distantly, there are layers of far-travelled fine mud that have settled. It is a new sedimentary system that is being added to every day, as the world's trawling fleets set out to work.

Muddy and sandy sea floors recover, after a fashion. There have even been claims that the ploughing of the sea floor is beneficial. Certainly, there are winners as well as losers: those species most resilient to physical disturbance, and to the choking effects of increased turbidity in the water, are ready to take advantage of the discomfiture of their more delicate neighbours.

Among the worst losers of all are complex, slow-growing submarine ecosystems. The tops of many deeply sunken mountains are—or used to be—often crowned with a delicate tracery of deep-water coral colonies that provide shelter for a variety of animals. In recent decades, the trawler fleets, in search of their particular prey, have simply smashed through the delicate coral growths, effectively reducing parts of the system to a mass of rubble. There is still an ecosystem of sorts among the dead and broken coral fragments, but one of generalist survivors—a far cry from the biological richness of the original deep-sea reefs. There are now controls in place to try to protect the remaining deep-water coral stands. How effective they will be remains to be seen.

Changing the Balance

Has there been anything in the Earth's past quite like the human-driven fisheries project? As an example of predation by one species—amazingly, a terrestrial one—on the ecosystems of both the open ocean (including surface, mid, and now increasingly deep waters) and of the sea floor, it is without compare: a blitzkrieg that is quite unique in the Earth's history.

Are there any partial analogues then? One struggles to think even of these. In the continuous arms race that marks the struggle for existence both on land and in the sea, there is usually a whole menagerie of winners. Each gains a small space where they keep some kind of grasp on the nearby resources, elbowing their direct competitors out of the way while, in adjacent parts of the food web, other species prosper. On land, for instance, until geologically recent times resources were shared between some 350 species of large land vertebrate. Then came humans, and they pushed most out of the way, or ate them (usually both). Now some two-thirds of vertebrate biomass is taken up by those favoured species (in a manner of speaking) that we maintain to eat—the cows, sheep, pigs, and so on, while another third is taken up by us, the eaters. Squeezed down now to about 5 per cent are the wild animals—the elephants, tigers, and rhinoceroses, and all of their wild kin.[98]

In the sea things are a little different. Up until recently, it has been largely a case of predation, pure and simple, removing mainly the top predators (whales, sharks, tuna, and so on) and a good deal of *their* prey, that is, the once-abundant shoals of fish—together of course with that thoroughly ploughed sea floor. We have not moved into the sea to take up permanent residence, although we like passing across it, and occasionally through it in submarines and bathyscaphes. But we have begun to farm the ocean systematically over the past few decades, largely by catching and grinding up types of marine fish that we do not like to eat very much and feeding them to those that we do, kept in large cages in the currently booming practice of aquaculture.

Through time there have been a good number of creatures that have moved from land into the resource-rich sea. Whales and dolphins had landlubber ancestors;[99] so did sea snakes and crocodiles, and ichthyosaurs and plesiosaurs before them. But in each case the transition to the sea was slow, via a progression of species that became successively better at paddling close to shore, to briefly swimming out a safe distance, to eventually cutting ties with land altogether. And in each case the final step did not mean attaining an easy and utter dominance. It simply meant sharing that ecospace, more or less successfully, both with earlier marine pioneers and with the lineages that had stayed true to the sea throughout—the sharks and large predatory fish.

In the Jurassic, for instance, the ichthyosaurs and plesiosaurs (many species of both) shared the position of top predator with the sharks and with fish such as the enormous *Leedsichthys*, over 15 metres long. There is usually some kind of coexistence here, if an uneasy one, among neighbouring tyrants. We humans have broken that power structure of contemporary marine predators. What is left is the small fry—that we are, too, harvesting as we fish down the food chain. For the oceans, it has been a unique form of reshaping—and one whose effects can only be compounded as we reshape the physical and chemical structure of the ocean. *That* process, as we shall see, has only just begun, and is so far mostly invisible to the casual glance—in temperate climes at least. But there's no way to avoid litter.

Litter

In 1997, Charles Moore sailed from Hawaii to Long Beach, California. It is a beautiful, calm stretch of ocean. The voyage took him a week—a week that should have been idyllic. His idyll was spoilt, though, by litter. There was not an hour in that week-long voyage when he did not see, bobbing in the sea, a bottle or a piece of plastic. Just how much litter was there in that stretch of sea, he wondered?

Moore returned, having persuaded some marine scientists to go with him. This time they trailed behind them a net that was designed to catch zooplankton. They also caught plastic. Lots of plastic. They counted the pieces of plastic that they had harvested, calculated how far they had travelled, factored in a figure for the size of the plankton net, and did some sums. There were, they calculated, 334,271 pieces of plastic floating in every square kilometre of that patch of ocean. For the Pacific this was a record. They did catch some zooplankton too but, even more astonishingly, the amount of zooplankton was one-sixth of that of the plastic. Charles Moore had discovered the Great Pacific Garbage Patch.

In some ways this is not surprising. Since the middle of the twentieth century, plastic, that wonderfully *convenient* material, has become a central part of all our lives. In the US alone it has become a trillion-dollar business employing more than a million people. Every year, 280 million tonnes (and rising) of this stuff is produced around the world. Something less than half of this is buried in landfill sites or recycled. The rest is used and then scattered to the four winds. Inevitably, a lot finds its way into the oceans.

In the oceans, the currents carry the plastic long distances. The current systems wrap around more slowly moving areas of the ocean, hundreds to thousands of kilometres across. These are the gyres that we have described in Chapter 5, and it is in these that the plastics become trapped and build up to the amounts that Charles Moore recorded. Gyres are common features in the oceans, and so there are other rubbish patches out there—in the South Pacific, in the Indian Ocean, in the Atlantic.

What happens to this plastic in the water? There are the large, obvious fragments—bottles, lengths of fishing line, bags. Eventually, though, these break down into small fragments, just a few centimetres or millimetres across. The denser fragments drift down to the sea floor. The rest simply stays afloat, being swept by currents into those enormous floating garbage masses, or washing up on beaches.

Angels' tears, the snorkelers and scuba divers call them—the delicate translucent scraps adding their own unsettling kind of beauty to the age-old patterns of fish and medusae, arrow-worms and lengths of seaweed. Are they, though, merely a highly visible but harmless addition made by humans to the natural world? Plastics, after all, are relatively inert (which is why they take so long to break down) and are designed to minimize toxicity. One can drink from a polystyrene cup without worrying about being poisoned by it.

Escapees from the human realm, the plastic can, at times, be anything but benign. Among the victims are the seabirds, seals, and turtles choked by plastic fragments or starved to death because their stomachs are full of an indigestible mass of human junk. Similarly, tiny, submillimetre-sized plastic fragments can be captured by filter-feeding plankton. Filter-feeders, large and small, are particularly vulnerable. They are unselective by nature, simply sweeping up suitably sized particles out of the water. If most of the particles in a water mass are made of plastic, then most of the food supply of a filter-feeding animal will therefore be of plastic too.

The dangers might reach down to tinier levels. Plastics are not quite inert. For a start, in the manufacturing process not all the individual chemical units—monomers—will react together to produce the polymer chains of the plastic, and these trapped, unreacted monomers are not chemically inert. Then, to make plastics more *plastic-like*—to make clingfilm cling, for instance—plasticizers can be added. In addition, certain plastics tend to absorb hazardous chemicals—polychlorinated biphenyls (PCBs), for example. Such plastic fragments can act as sources of slow-release toxins to the animals and birds that eat them.

These dangers, actual and potential, have led to calls for plastic waste to be classed as chemically hazardous (a status it does not enjoy at the moment), in the hope that it may help control the spread of plastics pollution. Current attempts at control are largely ineffective. There has been an international ban on the disposal of plastic waste by ships into the sea for decades—signed by 134 nations. Since then, the recorded levels of microplastic debris have risen, not fallen. Not all ships comply with international regulations, and in any event much of the plastic is washed in from the land (for which no international regulations are in place).

How bad can the problem get? The use of plastics is soaring. With business as usual, the amount of plastic produced by the world by the middle of this century will reach an estimated 33 billion tonnes.[100] That would be enough to wrap the whole Earth, land and sea, in clingfilm more than five times over. It will likely feel like that, too, unless either levels of production fall dramatically or global levels of recycling climb even more strikingly from their current 10 per cent or so. The plastic planet is growing ever more a reality. The plastic oceans somehow have an added poignancy.

Warming

A trawled sea floor can recover if left alone for a while. Fish populations can bounce back (although perhaps into another pattern), given enough respite. The world's plastic layer can gradually degrade into harmless and unrecognizable dross, or be buried by marine muds. However, some of the threats to the sea—although seemingly invisible or imperceptible—can alter the Earth's seascapes far more profoundly, and forever. Or, at least, forever as regards the prospects of recovery of the oceans for human use and enjoyment: for 10,000 human generations, say.

Three major threats are directly tied to humanity's current addiction to carbon-based fuels: warming, acidification, and oxygen deprivation. It is still astonishing to think that altering the proportion of what is after all a trace gas—carbon dioxide only comprises a little over one-tenth of one-third of 1 per cent of the atmosphere—can have such far-reaching consequences. Astonishing, and of course to many still unbelievable. But the natural world is complicated and often a little counter-intuitive, and the oceans are, in some ways, more vulnerable to change than the land.

Emissions of carbon from their ancient storehouses in the ground—of coal, oil, and gas—into the air are broadly known. They currently stand at about 10 billion tonnes per year (just a few years ago one was talking about 7 billion tonnes a year, but then that is the nature of progress). That translates into about 30 billion tonnes of carbon dioxide, because two oxygen atoms (atomic weight ~16) are added to each carbon atom (atomic weight ~12). That's just a figure. We can't absorb figures in any real fashion, or at least most of us can't.

A kilogram of carbon dioxide, if in solid 'dry ice' form, makes up a cube with each edge a little over 8.5 centimetres long, like one of the larger children's play bricks. In gas form, at normal atmospheric temperature and pressure, it swells out to something over five hundred litres, which is about the volume of a *very* large family refrigerator. A transatlantic flight in a 747 releases over 200 tonnes of carbon dioxide, or 200,000 refrigerators full. As the *New York Times* commentator Andrew Revkin once said, if carbon dioxide was pink, we would be *very* aware of what we are producing (and if it was smelly, we would have turned to another energy source forthwith).

What about other sources? Volcanoes are huge carbon dioxide emitters, surely? They are, but not on the gigantic scale at which humans are now operating. The 1990 eruption of Mount Pinatubo in the Philippines was the largest of the past century, explosively expelling about 10 cubic kilometres of magma which spread over the surrounding countryside as ash and pumice fragments. That eruption—a catastrophic, gas-rich one—released something like 42 million tonnes of carbon dioxide, which is certainly large but does not even appear as a blip on the rising curve of carbon dioxide concentrations.

There are larger movements of carbon dioxide out in the natural environment. Each year some 60 billion tonnes are fixed as plant matter, and an equivalent amount rots and decays to release the same amount into the atmosphere. The process is geographically asymmetrical. There is more land, and therefore more plant matter in the northern hemisphere, and so the Earth 'breathes' annually; these 'breaths' are seen as the annual rises and falls of carbon dioxide in the rising trend. Further, each year some 90 billion tonnes dissolves into the oceans and—at least until a couple of centuries ago—the same amount was exhaled. This is all part of a grand, finely functioning, and delicately poised global carbon cycle. Despite these huge inflows and outflows of carbon—and the even larger reservoirs from which they are derived (the oceans, for instance, store within them over 45 thousand billion tonnes of carbon)—there has been a quite extraordinary long-term stability of atmospheric carbon dioxide for the thousands of years before the Industrial Revolution that is nothing short of amazing (or at least we find it so). Perhaps it was not quite a natural stability, given the possible effects of early agriculture, but nevertheless it shows how finely regulated the land/ocean carbon cycle normally is, when left to itself.

We know this because air—pure, pristine air from the past—can be fossilized. Not in normal rock—at least, not as far as we know—but in ice. The layers of snow that form every year over the vast ancient icecaps of Greenland and Antarctica (and also the somewhat more short-lived icecaps of the Himalayas, Andes, and other mountain ranges) are initially light and fluffy, being full of air. As more layers of snow fall on them they become compressed, and much of the air is squeezed out—but not all. As fluffy snow converts into firm snow and then into ice, tens to hundreds of metres below the icecap surface, what was a continuous network of air passages closes off into myriad tiny air bubbles that, from then on (and no matter how much more compression takes place) are simply locked into the ice.

Drill into that ice, as teams of scientists have done in both Greenland and Antarctica (and in other ice masses too), and one obtains a record of air that reaches back almost 870,000 years (with an aim to try to obtain air a million years old) from beneath the centre of Antarctica. That is possible—but about the limit, because the ice is not permanent. It flows under its own weight out to the edge of the continent to break into icebergs, which then melt, releasing (with faintly audible 'pops') air from the time of the mammoths and Neanderthals back into the atmosphere.

The fossil record of air reveals an even more remarkable stability—or at least a stable oscillation between two modes. About 180 parts per million (ppm) in glacial phases, and about 280 ppm in interglacial phases. Even at these longer timescales, the Earth's carbon regulation systems is—again, when left to itself—a very finely regulated machine.

Now of course that machine has been pushed, with unprecedented speed, into a new gear. There is an iconic curve that shows the climb of carbon dioxide in the atmosphere since 1958, painstakingly and determinedly produced by the late, remarkable scientist Charles Keeling (a singularly stubborn man, to carry on taking measurements, year by year, that others originally told him were valueless). It starts at under 320 ppm and now, not much more than half a century on, stands at 400 ppm. To see the effect in proper—that is properly geological—context, though, it needs to be grafted on to the almost-million-year ice core record. At this scale it is a cliff: a straight vertical line terminating now in an atmosphere the Earth last experienced in the Pliocene Epoch, before the Quaternary Ice Ages. It is, very likely, the most rapid large-scale release of carbon dioxide the Earth has ever experienced.

Another century on and where will we be? We are in the realm of the various scenarios produced by the Inter-Governmental Panel on Climate Change and others—and, for now, are at the higher end of the range of projections. Currently, each year adds another 3 ppm. If that continues, the Earth will see 700 ppm by the end of the century. That will be an atmosphere more like that at the time of the dinosaurs than that of today.

The Earth now has a thicker blanket, and it is perceptibly warming. It warmed by the order of a degree in the last century, globally, albeit with ups and downs. Much of the extra warmth retained by the Earth has gone into the oceans (which has stopped air temperatures rising so quickly), and this has probably caused about half of the 30 centimetres of global sea level rise that has taken place over the past century. Given how long it takes the ocean waters to circulate and mix, much of this extra warmth is still at the surface. It will take centuries for it to slowly spread through into deep waters, and begin to raise them above their current frigid state. The near-surface warming is most strikingly visible around the poles, partly because these are the parts of the world that are warming most quickly, and partly because the sea ice of the polar regions is a highly tangible indicator of temperature change.

The declines in Arctic sea ice over the past couple of decades caught everyone by surprise, not least the oceanographers: the computer models that they had painstakingly built to try to peer into the warmer world to come had suggested that a good cover of Arctic ice should persist into the end of this century. The figures that then emerged from the year-by-year satellite surveys alarmed them. Between 1979, when the satellite surveys began, and 2012, the area of sea ice remaining at the end of the summer melt had almost halved. The figures for volume were yet more shocking: measurements of ice thickness showed that there was only a quarter of the Arctic ice in 2012 that there had been in 1979. The ice that remained was much thinner—much of it only a skim one season old, by contrast with the abundance of thick, multi-year ice that there used to be.

The decline was not smooth. Some years showed more rapid declines than others. Some, indeed, showed increases (there has been one, as we write[101])—while yet others saw steep plunges in ice areas and volumes. But the overall trend is clear. Could it be, though, a longer-term blip? We know that climate and ice amount can vary over timescales from years to millions of years. As ever, in attempting to understand the Earth's behaviour, we go back to the past to see whether the seemingly extraordinary is (geologically) an everyday occurrence, and thus whether today's melt might only be a prelude to the regrowth of ice to normal levels in the next few decades.

What does the long-term history of Arctic ice tell us? An array of evidence has been painstakingly collected from the Arctic regions. Some evidence comes from the present-day shorelines of landmasses—Greenland, Arctic Canada and Alaska, Spitzbergen, and elsewhere. Some has been collected, as borehole cores of deep-sea sediment layers, from the floors of the Arctic Ocean, where icebreakers have accompanied the drilling-ship to allow safe passage. Some evidence has been collected from the ice itself, as annual snow-layers have, year by year, piled up on the landmasses adjoining the Arctic Ocean and then been compressed and hardened into ancient ice.

The story is clear enough.[102] Arctic sea ice is ancient. There is evidence of perennial ice from 13 million years ago, when grains of iron oxide from Siberia began to appear within the sediments accumulating on the Arctic sea floor. This is a clue that the sea ice had changed from seasonal to perennial in the Arctic, for these specific mineral grains could only be plausibly carried so far by being rafted on persistent multi-year ice. Since that time, this surface ice layer has been highly sensitive to climate change, waxing and waning as the Earth has successively cooled and warmed. During the coldest phases of the Ice Ages of the last 3 million years it may have become a solid armour, perhaps reaching several hundred metres thick. When the warm interglacials came, it shrank back, at times virtually disappearing altogether—as it seems to have done in the warmest part of the last interglacial phase, some 125,000 years ago.

The history of Arctic ice in the current, Holocene, interglacial phase may be tracked, for instance, by searching for bones of bowhead whales along Arctic coastlines, because these follow the sea-ice front to escape their main predators, the killer whales (although it is a risky tactic, for whole schools of bowheads can be trapped and killed by enveloping ice if they linger too long as winter sets in). Such evidence shows that the early part of this interglacial was a little warmer than present in the Arctic, with reduced ice cover. Since then, the ice (and the bowhead whales, and the Inuit who hunted them) has been spreading back southwards.

This trend has now been sharply reversed, and the current rapid shrinking of the ice has no parallel in the last few thousand years. It is a striking anomaly, one of the clearest indications of human modification of climate—and it includes a strong reinforcing mechanism, for open water absorbs more of the Sun's heat than reflective ice, ratcheting temperatures yet higher. The computer models are being rapidly revised, and we can expect an ice-free Arctic Ocean in summer by mid-century, perhaps earlier.

As carbon dioxide levels continue their inexorable upward trend, it is becoming more likely that the Arctic ice, once lost, will not return for thousands of years, and will be a crucial part of a permanently (as far as we are concerned) warmer ocean system. What will be the effects?

Life in a Warmer Ocean

The warming of the last century has produced changes in life on land as plants and animals have migrated to follow the climate belt that best suits them. On land, temperatures naturally vary from day to night and from day to day, and living plants and animals must adapt to such conditions. The ocean waters, though, because of their bulk and their capacity to store heat, are much less prone to such short-term fluctuations. Marine organisms, therefore, are used to living in more uniform temperature conditions, and are more sensitive to changes in temperature. The results of even slight temperature changes can look scary.

Take, for example, the television pictures beamed, in 2010, up from a remotely operated vehicle that was trundling along the sea floor off the Antarctic Peninsula to its mother ship, the research icebreaker R.V. *Nathaniel Palmer*. The crew huddled around the screen were perplexed, for something strange was happening down there, almost a kilometre and a half beneath them. They were used to seeing the typical life forms of the Antarctic sea floor—sea lilies, brittle stars, and peculiar forms of sea cucumber, stumping around the sea floor on short legs. These are archaic faunas, more like throwbacks to the far-off Palaeozoic age than the typical denizens of the modern deep sea. But there was a bare patch there. Then, the watching scientists saw why. A red crab, its legs spanning the diameter of a large dinner-plate, walked into view. Then another. And another. They were looking for food, and the long-isolated creatures of the Antarctic sea floor were easy prey.

The crabs had arrived with the warmer water that is now impinging ever deeper into the normally frigid Antarctic bottom regions.[103] It is the beginning of a classic biological invasion. If it continues, as seems likely, it will devastate the pristine ecology of the deep polar regions quite as effectively as rats, pigs, and goats have devastated the original ecology of most islands in the world. Such an invasion is part of the warming-related biological change which has now been documented around the world—but that particularly affects, for now, the more sensitive high latitudes (low latitudes are not entirely unaffected, as we will see with corals).

There are more subtle effects of changing temperature too. As the sea surface warms it becomes less dense—and so harder to mix in with the colder, denser, nutrient-rich waters below. The sea becomes more stratified. This makes life harder for the microscopic plankton that form the base of the food chain, because as oceans become more stratified they become more nutrient-starved.

A very simple device invented by an Italian priest suggests that this might already be beginning. In 1865, Father Petro Angelo Secchi (1818–1878) was asked to come to the aid of the Papal Navy, who wanted more information on the clarity of the Mediterranean waters that their vessels sailed through. The device Father Secchi designed for them couldn't be simpler. It is a white disc, which is lowered into the water until it is no longer visible. The more plankton in the water, the sooner the disc is lost from sight. For Secchi this was child's play. A Jesuit priest, he was director of the observatory at the Pontifical Gregorian University in Rome. He invented spectrographs to analyse the spectra of the Sun and of other stars—and using this evidence was one of the first scientists to be able to say that the Sun really *is* a star. He discovered several comets too, and, more curiously, was the first person to refer to 'canals' on the surface of Mars. His interest in oceanography, though, was serious enough; he met and corresponded extensively with one of its then leading figures, Matthew Fontaine Maury (see Chapter 3).

Secchi discs soon became a standard measure taken on board ship and, until the advent of satellites, formed the only systematic measure of plankton abundance in the oceans. Recently, Secchi disc measurements made throughout the world oceans for the past century have been analysed. It was an enormous task—but it was worth it. The measurements fluctuated from year to year and from place to place, as might be expected—but a clear overall trend appeared. Over the past century global ocean waters have become clearer and hence less plankton-rich.[104] The total decline is 14 per cent—a surprisingly large figure given that the world's warming is only just beginning. Plankton makes up the base of the food chain, and so the totality of open-ocean life (regardless of what form it takes, and how humans might affect its proportions by fishing) has already declined significantly. As the oceans continue to warm and to become more stratified over the coming decades and centuries, one might expect open-water ecosystems to shrink further. This is food for thought, not least in a world where a growing human population will doubtless seek to take yet more fish from the seas.

The Oxygen Deficit

As the oceans become more stratified, and it becomes harder to extract nutrients from the depths to the sunlit surface levels, so it will also become harder to mix in oxygen from the surface into deeper levels, especially into the oxygen minimum zones that are present at intermediate depths, above those waters that are stirred and supplied with oxygen by deep-water current systems (see Chapter 5). Reductions in oxygen content have already been observed, and are predicted to spread farther as the world continues to warm. This has serious implications, for fish need a good oxygen supply; as this declines, one predicted response is that fish will simply get smaller.[105] There are precedents for such change in the coincidence of ocean anoxia with greenhouse states of the geological past (see Chapter 5). So far, the effect is subtle. Closer to shore, less subtle and more peculiarly human modification of the ocean is apparent. Here, the dead zones are spreading.

The term 'dead zone' is a striking one, but in this case the term reflects contemporary reality. The world today is crowded, and 7 billion people are currently only being kept alive because we have invented means to supercharge soil fertility to boost crop yields far above what would naturally grow. One of the sources of this fertility is additions of phosphate[106] which, in the past, was supplied by manure, by Chilean guano, by crushed animal bone from the slaughterhouses, by fossilized dinosaur bone (this was a thriving and very profitable trade in England for several decades), and even human bone. In the nineteenth century, gangs of English labourers roamed the battlefields of Europe, scouring them of the bones of millions of dead soldiers to transport back, grind up, and spread on the fields—to the horror of delicately natured observers. In perhaps the most bizarre example, a large nineteenth-century find of ancient Egyptian mummified cats from Egypt was sacrificed (one more time, perhaps) to help the crops grow. It is a measure of how desperately sought this nutrient was—and still is, for there are fears that our current supply, from phosphate rock strata, will, sometime soon, fail to meet demand.

The other main means to supercharge our soils, nitrogen, is in more abundant supply as it makes up most of the atmosphere. But fixing it into a form that can be absorbed by plants is trickier. Certain bacteria do this naturally, but too slowly for our current insatiable demand. When Fritz Haber (who we met in Chapter 4, seeking gold in seawater), together with Carl Bosch, invented a means of artificially creating nitrogen-based fertilizers from atmospheric nitrogen, the stage was set for the subsequent more-than-tripling of human population. The associated perturbation of the global nitrogen cycle has been said to be the greatest since Precambrian times.[107] Since the mid-twentieth century the amount of reactive nitrogen at the Earth's surface has roughly doubled. This means that one species—the human species—is putting as much nitrogen into the Earth system as the rest of biology put together.

Not all of these fertilizers go where they are intended, into the crops. Part is washed from the fields into streams and rivers, and then into the sea. There they stimulate plankton blooms that, on dying, fall to the sea bottom. In those regions where fertilizer inwash is high and the sea's circulation is naturally sluggish, the decaying plankton can use up all the oxygen. The result is suffocation for all the animals living on, in, or just above the sea floor. There are now about 400 such dead zones in coastal areas around the world, covering in total about a quarter of a million square kilometres, of which the most notorious are in the Gulf of Mexico, the Baltic Sea, and Chesapeake Bay on the eastern US seaboard.[108] It is generally an annual phenomenon—mostly, winter storms stir in enough oxygen to allow recolonization of the sea floor by a new generation of animals—animals that are fated only to survive until the next summer; the ecology of the dead zones is now restricted to those organisms that can tolerate such short and interrupted life cycles.

It is hard to think of these new marine dead zones as anything other than horrific. But the bottom of the sea is rarely in people's thoughts, and this significant change to the world's marine ecosystems is mostly out of sight and out of mind. Nevertheless, as ever more fertilizers will be needed to grow the crops to feed an expanding global population, the dead zones seem set to spread ever more widely. These localized zones can be added to the more widespread oxygen depletion of the warmer, more stratified seas of the near future. For the myriad organisms for whom the oceans are home, it will simply be harder to breathe.

Acid Oceans

For a long time, the fact that the oceans are absorbing a good deal of the carbon dioxide that we pump out into the atmosphere was seen as an unmitigated benefit. After all, the more that there is stored in the oceans, the less that remains in the atmosphere to trap heat and warm the Earth. And there is the 45,000 billion tonnes of carbon that is stored in the ocean in one form or another—bound within organic matter, or as carbonate and bicarbonate molecules, or simply as dissolved molecules of carbon dioxide. This is an amount that is vastly greater than the ~550 billion tonnes of carbon that were stored in the atmosphere before humans took a hand in the process, and the 800 billion tonnes that the atmosphere holds today. A little more would not make much difference, surely?

There have even been ideas of compressing carbon dioxide from power stations and pumping it down to the deep ocean floors. There, at the near-freezing temperatures and high pressures of that realm, the compressed carbon dioxide would stay as a dense liquid for a long time, hugging the ocean floor and only slowly being dissolved into the waters above. Only a few obscure deep-sea organisms would suffer, it was said, while the rest of us would benefit. In recent years, though, people have gone quiet about this idea.

In hindsight, the idea that dissolving more carbon dioxide in the oceans may have a dark side was extraordinarily late in being widely realized. From the late 1970s, oceanographers raised concerns over possible unpleasant unintended consequences of humanity's giant experiment with carbon dioxide. The chemistry was, after all, straightforward—or at least *moderately* straightforward. The carbon dioxide dissolved into the oceans will combine with water to form carbonic acid, which releases the business end of its acid nature—hydrogen ions—into the water. The pH drops, changing the balance of the carbonate and bicarbonate ions already dissolved in the water, shifting the balance away from carbonate towards bicarbonate. It is the carbonate in the ocean waters that is the feedstock for those organisms—molluscs, corals, foraminifera, sea butterflies—that build skeletons of calcium carbonate. The less dissolved carbonate there is, the more difficult it is for these organisms to build their skeletons. But how big is this problem?

The answer is that it is absolutely enormous. The statement of the scale of the problem that made everybody sit up was published in 2003, in a 'brief communication' to the journal *Science* by Ken Caldeira and Michel Wickett, scientists at the Lawrence Livermore Laboratory in California.[109] They devised a computer model that tracked the release of carbon dioxide from the present into the future under a 'business as usual' scenario—that is, assuming that we will burn our way through most of our fossil fuel resources over the next few centuries. They modelled the carbon emissions from the air into the oceans, based on the way that carbon dioxide partitions between atmosphere and ocean (which is moderately straightforward) and the way it then travels through the oceans. This second factor is a little trickier, involving knowledge of how ocean waters circulate to carry the carbon dioxide, but in essence the carbon dioxide will build up first in shallow water and then be carried into deep water. Let this experiment run, they said, and the pH of the water will drop by as much as 0.77 of a pH unit. To put that into context, this would be a greater change than any seen in the last 300 million years at least—with the possible exception, they added, of 'rare, catastrophic events in Earth's history'.

Already, the pH of seawater is one-tenth of a percentage point lower than it was before humans started burning fossils fuels in earnest at the beginning of the Industrial Revolution. That may not sound very much: it is a drop from a pH of 8.4 to 8.3, and the water is still well

above the neutral level (of pH 7.0)—so it is not 'a
basic). Nevertheless, for the creatures that live in the
amount of change from the status quo that matters. Th
logarithmic, not linear, and so a change from 8.4 to 8.3 mea
30 per cent more hydrogen ions in the water. It is also roughly
amount of change in ocean carbon chemistry that took place between
glacial and interglacial phases of the Ice Ages—only we are heading beyond interglacial levels into uncharted territory.

How much of a problem is this for sea life? After the relative simplicity of the chemistry, the biological effects prove more complicated. The building of calcium carbonate skeletons takes place within biological tissues, and so the chemical micro-environment (or perhaps more accurately the nano-environment) is controlled not just by the chemistry of the seawater around them, but by the ability of various organisms to control the chemistry within their tissues, and thus to counteract changes in external circumstances.

Hence, biologists have been scrambling in recent years to find out what rising acidity levels would do to different types of marine creature by keeping them in tanks in which different amounts of carbon dioxide have been dissolved, or by observing what happens when seawater is naturally made more acid—where volcanic gases escape from submarine volcanic vents for instance. These studies show that rising acidities can interfere with the formation of calcium carbonate skeletons, but that different types of organism have different tolerances.

Let us take a couple of examples. Coral reefs are being observed particularly closely because of their enormous ecological importance. They are, like the tropical rain forests on land, storehouses of much of the oceans' biodiversity, some 25 per cent of all marine species being crammed into about a tenth of 1 per cent of ocean area. They are remarkable, counter-intuitive biological systems. Their diversity is predicated upon the elaborate architectural framework built by the

coral colonies, and needs *low-*
gs (because the coral systems are
trients there are). Coral reefs are
els, and have suffered as human-
e washed into them from the land;
se of the corals. Corals are also sen-
o hot, the coral animals expel from
lagellates (a type of single-celled pro-
tist) tha grow. The corals then lose their vivid
colours: they 'bleach', and ...lly die. In recent years there have been major bleaching events, killing off large areas of coral during tropical heatwaves (Fig. 17). Corals are also sensitive to disturbance, both physical (dynamiting, building, dredging) and biological (human hunting

FIG. 17. Bleached coral, Great Keppel Island, eastern Australia, 2006. Intertidal corals were badly damaged by excessively warm conditions.

of ecologically important reef fish). In short, coral reef systems worldwide today are struggling (some are already dead), and the growing acidification threatens to be the coup de grâce (see Plate 4).

The latest studies suggest that at atmospheric carbon dioxide levels of somewhere near 550 ppm the coral animals will not be able to maintain a positive balance of calcium carbonate formation. From around that point (and it will vary between different species, and in different settings), the coral reefs will stop growing and start to shrink back. At current rates of carbon emissions, this particular tipping point for reefs will occur around the middle part of this century. Small wonder that some marine biologists have referred to coral reefs as zombie ecosystems: still alive, if increasingly in poor health, but doomed. The Earth, then, would go through another of the events that geologists term a 'reef gap', when these magnificent, diverse structures disappear from the world. The last reef gap was 55 million years ago—and was also associated with an ancient global warming and marine acidification event, termed the Paleocene-Eocene Thermal Maximum. It took millions of years for the reef systems to recover.

It is not just the corals. Other creatures are already showing signs of being eaten away, just as a result of the 0.1 pH increase so far. Pteropods—otherwise known as sea butterflies—are planktonic molluscs that secrete delicate and rather beautiful shells of calcium carbonate. They are also rock builders, their shells falling on to some areas of the deep-sea floor in such amounts that they build sedimentary layers. Pteropod skeletons have lost weight in recent years.[110] They too look to be in danger of declining, perhaps vanishing, this century.

Acidification does not only affect skeleton building. If the pH of seawater drops to the kind of levels envisaged, then that interferes with physiological processes such as respiration in fish, too.

The Future

We have focused on only a few of the current, unprecedented and (currently) accelerating changes to the oceans. There are others: pollution—hydrocarbons, heavy metals, human-made chemicals of all kinds—that flood into the sea; noise pollution—the din from our boats—that hinders communication between marine mammals; the change in sediment patterns caused by damming rivers and coastal construction works, and the tearing up of mangroves to build shrimp farms; and species invasions that are not climate-related, but result from the transport of species by humans. (In the sea, a very effective means of transporting species has been the widespread use of ballast tanks, by which sea floor sediment, complete with living creatures, is simply scraped up in one part of the ocean when needed to make up weight, and then dumped elsewhere.) These and yet more contemporary changes have been eloquently summarized in books devoted to this subject.[111]

The situation is not quite hopeless. Measures such as marine reserves have been put into place, and these can be effective, at least as shelter from human predation. The single most dangerous by-product of human civilization in this respect (and in others) is carbon dioxide. The single most effective measure to bring it under control is likely to be an effective carbon tax, given the central position of money in our lives and the failure of the touted alternative, the carbon market, to make a scrap of difference to rising carbon dioxide levels. A carbon tax could be linked to a stable regulatory framework to encourage business investment into, and incremental improvement of, non-carbon-based energy sources.[112] Substantial decarbonization of the world this century is possible (and indeed can be profitable, because it entails growing a very large industry). Carried out quickly enough, it might even confound the pessimists and preserve some part of the ocean's current biological riches, not least some functioning coral reef systems. Let us hope so.

However, we should separate what we would like to see (a global economy shifted to a pattern that can preserve the oceans in something like the form they are in) from what currently seems most likely. In the business-as-usual political realities of our human economy there currently seems not the faintest chance of stopping carbon emissions over many decades, let alone overnight. Business is, indeed, currently accelerating, as a population now exceeding 7 billion, and heading for 9 billion by mid-century, strives to live as best it can.

The most likely end result is that the diverse, beautiful ecological systems still dominated by reef corals and fish will be replaced by 'slime-rock' systems dominated by algal and microbial mats and jellyfish. Indeed, the marine biologist Daniel Pauly has coined the term 'Myxocene' (derived from the word for slime) for such a likely future ocean state. Using other new terminology, it will be part of the developing Anthropocene Epoch.[113]

The oceans will become warmer, more oxygen-poor, and, over most of its surface, more oxygen-starved. Warm-water species will, over the coming centuries, expand out towards the poles. Food webs will break down and reform into new, simpler, more species-poor combinations. There will, more likely than not, be a mass extinction event to perhaps rival some of the great catastrophes of the geological past.

On a cosmic scale this is important, because the Earth is a cosmically rare jewel. On a cosmic scale too, the Earth will recover, as it has done following the great perturbations of the past—once (or if) humans and their unique planet-altering capacities have themselves become extinct. The recovery process normally takes several million years, but it will be a new ocean state (particularly biologically) that will appear, and not a return of the old one. The oceanic riches to come, in the far geological future, we might guess at just for fun, and we have to consider also the last hurrah of oceans on this planet and what comes after, for the story of the Earth and its seas is not destined to be never-ending.

8

The End of Earthly Oceans

When will the world end? In some religions apocalypse is just around the corner, and we have to prepare for it, and for the next and better world that awaits. In others—Hinduism and Buddhism, for instance—the timescale is unutterably vaster, and equates to many billions, and perhaps trillions, of years. For the Comte de Buffon, writing the world's first scientific Earth history text in pre-Revolutionary France, the world was set to freeze in a few thousand years, with the evil day a little forestalled by humanity's burning of coal. A little later, James Hutton, Scottish landowner and savant, discovered the abyss of deep geological time, in seeing evidence in strata of mountain belts first built and then worn down to their roots. Famously, he mirrored the Buddhist and Hindu timescales in seeing, for the Earth, no vestige of a beginning or prospect of an end.

But the universe, we now know, began 13.8 billion years ago in the Big Bang, while our solar system, and the Earth, are just 4.6 billion years old. On the timescale of modern science, what prospects do we have? Will oceans last until the end of the Earth? Or will the Earth's old age be parched and arid? If so, it may become a planet that we might not recognize, were we to take a trip there in that old time-travelling telephone box. Indeed, we might be well advised not to open the doors.

The timescale we are now thinking of is one that ranges from hundreds of millions of years to billions of years. For us humans—for any humans—that is far beyond any practical concern. We will have quite enough challenges to deal with over the next few decades and centuries simply to keep ourselves going as a globally successful species, or perhaps even as *any* kind of surviving species. The next few pages, therefore, serve only to fuel our curiosity about the Earth's future at a scale at which, *almost* certainly, it will be of no practical relevance to us whatsoever.

Still, we have an attachment to our home planet that goes beyond anything that one might put within the categories of our own self-interest, or that of our children, or a love of nature, or a marvelling at the beauty of rain forests and coral reefs. It is the idea that there is a world—the only one that we know of—on which there are living beings: orchids, kelp, sponges, amoebae, sea anemones, squirrels, krill, marlin—it almost doesn't matter *which* living organisms, so long as they are there. The thought that the Earth might one day be lifeless brings about a feeling of emptiness—a spiritual emptiness that might only be assuaged a little if the next generation of space telescopes detects, on some far-off planet, the signature of complex chemistry sufficiently out of equilibrium to show that life is out there, and that we are not alone.

The question of how long the Earth will have a functional biosphere is tied closely to the question of how long the oceans will stay in something like the form they are in today, for they—through the global hydrological cycle—regulate life on land almost as much as they shelter the living assemblages of the seas.

How long, therefore, have the oceans got? A little less than we might like, it turns out—for the sake of our existential yearnings at least. While the Earth as a rocky planet is not far off its midpoint, its biosphere is approaching old age. Why that might be we will examine below, but in any case the world, with us, might really be changing rather more fundamentally than the brief ecological wrecking spree we described in Chapter 7. Have we, perhaps, brought on a step change that will affect both land and ocean forever?

Breaking the Rules

The Earth is, in many ways, controlled by the film of life at its surface (as we saw in Chapter 6). Without that film, and the way it helps to regulate element cycles and climate, it is an open question as to whether our planet would have retained its oceans for over 4 billion years.

We know that those long-lived oceans are currently undergoing something of an ecological crisis that may, in a few brief centuries, come to rank with those of the great perturbations of the Earth's geological past. If that projected crisis takes hold, with rapid ocean warming, acidification, and oxygen deprivation, then the life of the oceans will change too, as ecosystems reassemble themselves within a mass extinction event.

At the moment all this seems more likely than not, given the trajectory and momentum of global change today. At their simplest, the man-made changes have a good deal in common with those triggered in the past by great volcanic outbursts or by rare, cataclysmic meteorite impacts—those that we now place at the boundaries between the Permian and Triassic periods, for instance, or between the Cretaceous and the Tertiary.

What comes after? After the great convulsions of the past, the great stabilizing processes of our planet swung into action: the removal of excess carbon dioxide from the atmosphere by its conversion, ultimately, into limestone rock and buried hydrocarbons; or the neutralization of other toxins (fluorine, sulphuric acid) by reaction with rocks or water. This kind of self-cleaning we know can take a few hundred thousand years. That was the timescale of previous 'hyperthermals' such as the burst of Toarcian warming in the Jurassic, or the Paleocene-Eocene Thermal Maximum. Planetary biology, after a serious mass extinction event, takes longer to build itself back up into the complexity and diversity it formerly enjoyed—albeit into an array of mostly different species. This healing—and transformation—of life has, in the past, taken a few million years, until the biosphere is rebuilt into its new pattern.

As we peer dimly into the future, the simplest scenario is to conceive of some future Earth, with future oceans, slowly evolving into a new natural pattern, once the remarkable pressures generated by humans have been lifted. This is assuming, of course, that our species will die out. All previous species on Earth have become extinct; few species exist for longer than a few million years—and our own species, we are fully aware, is putting itself at risk by rapidly undercutting its own planetary life support systems. With humans out of the picture, we might then simply use our understanding of both Earth history and of planetary systems to chart a way into the far future. This, indeed, is the path we will generally follow in this chapter.

But what if humans survived? Humans are resilient, manipulative, and ingenious as well as ecologically reckless, and they may be constructing something quite new on this planet that might interact with and affect oceans and land alike, and that, once in place, might prove robust and self-sustaining. The technosphere, it has been called.

The technosphere is the idea of Peter Haff, a professor of geology and civil engineering at Duke University in North Carolina, who has been thinking through the effects of technology at the most fundamental of scales. The technosphere may be compared with the century-old idea of the biosphere, which was, to all intents and purposes, the brainchild of Vladimir Vernadsky, a Russian scientist who performed the extraordinary feat of surviving through both Tsarist and

Stalinist times while arguing forcefully for academic freedom with both of these regimes. The biosphere, he said, was not just the assemblage of all animals and plants at the Earth's surface. It was all of these linked together, and linked too with their rocky and watery substrate, to store and transform the Sun's energy. The biosphere, he said, is a system that regulates conditions on the face of the Earth.

In 1920s Paris, Vernadsky, together with the French scientists and philosophers Pierre Teilhard de Chardin and Éduard le Roy, added the concept of a noosphere, a sphere of human thought around the world which was itself beginning to control processes at the Earth's surface. Peter Haff has taken that idea and turned it on its head.[114] We are not so much in control, he says: rather, the system that we have created is now evolving according to its own dynamics.

The 7 billion humans on Earth today are kept alive only through the continuous action of an enormous, globally interlinked system of transport and communication, metabolized by the use of vast amounts of energy (mainly from fossil fuels), and controlled by our bureaucracies. It produces and distributes all the materials that we need for food, shelter, and everything else. Without it, most of us would not be alive—and therefore we are forced to keep it going. Viewed from the inside, we create and control it. Viewed from the outside however, Peter Haff argues, it has developed an existence of its own, and we humans are entrained within it as component parts (in that we have currently no realistic alternative). With its own novel material character and dynamics, it may be regarded as an emergent system—one that, like all systems, appropriates matter and energy from its surroundings for its own existence and development.

The technosphere is developing and evolving at extraordinary speed, and now its component parts (computers and mobile phones, for example) are virtually unrecognizable from one generation to another. By linking, ever more closely, its biological components (people and the ideas that they generate) it is accelerating its own development. The technosphere, suggests Haff, is the force that is changing the Earth—by damming rivers, hoovering the fish out of the oceans, and hugely modifying natural elemental cycles.

Is it inherently robust though, in the sense that the biosphere is robust and has never lost its grip on the planet, no matter what catastrophes the Earth has endured in its history? It is much too early to say. A major threat to the technosphere currently is that it has not yet learned to recycle its own waste products: think of those plastics building up in the ocean gyres for instance, or that invisible but more sinister phenomenon of the carbon dioxide building up in the atmosphere and oceans. The biosphere, after billions of years of practice, has become very good at recycling. If the technosphere is to survive for anything approaching geological timescales as a system it has to alter its functioning, before it chokes itself and its human components on its own waste products.

The novel options that may be opening to this planet allow no sensible prediction. They may well, of course, turn out to be mirages, quickly built and then quickly destroyed by human hubris. Then we may safely return here to the realm of a natural Earth. And here we do have some guidelines.

The New Biology

As the dust settled from the impact of the 10-kilometre-diameter meteorite that ploughed into the Yucatán peninsula 65 million years ago, much of the world's ecosystem collapsed. It may well have been in poor shape previously, because—as planetary luck would have it—the meteorite arrived just as the Earth was suffering the toxic effects of a particularly intense episode of volcanic eruptions on the other side of the planet, on what is now the Deccan region of the Indian subcontinent.

The resulting extinction, on land, of the dinosaurs—of the *non-avian* dinosaurs, that is—is now as much a part of popular as of scientific culture. But much else died out as well, especially in the sea. Of the impressively monstrous creatures, the mosasaurs and plesiosaurs vanished. Among the small fry, the ammonites and belemnites were lost, and most species of brachiopods (lamp shells). Among the creatures that really count, those at the base of the food chain, many species of planktonic algae disappeared too.

It was a catastrophe, a body blow to the biological fabric of the Earth. But, as with all such mass clearings-out of species, there are opportunities too for those that survive. The following several millions of years saw extraordinary evolutionary radiations. On land, famously, the mammals took over the upper echelons of the food web. In the sea the mammals also prospered. The whales, originally those wolf-like creatures paddling along the shoreline, dipped more than a toe in the water to take over much of the territory formerly occupied by the marine saurians. The brachiopods never really recovered (they are still clinging on, just), while the resurgent bivalve molluscs and gastropods prospered. From among the microscopic forms, giants developed, such as the nummulites; single-celled foraminiferan protists they may have been, but nevertheless they built centimetre-sized shells in such numbers that Egyptian kings would find the resulting rock the material of choice to build their pyramids with.

But who could have predicted, as a thoughtful spectator amid the rubble of the dying Cretaceous world, what would have emerged? That there would arise such creatures as penguins and walruses, sea horses, nummulites, and narwhals? Evolution, especially as regards its more baroque creations, is not directly predictable.

One can have fun trying, though. Some years ago the science writer and illustrator Dougal Dixon wrote and illustrated the book *After Man*.[115] This zoology of the far future is set some 50 million years from now. Life has recovered from its encounter with humans, and gone on to evolve an impressive array of new species. There is, for instance, the woolly gigantelope (*Megalodorcas borealis*) roaming the tundra, with its shaggy mane, fatty hump (to help survive the bitter winters), and single long forward-pointing horn. Roaming the grasslands of his new world are herds of rabbuck—more or less deer-like in shape, although without the horn and with noticeably long ears. That is the giveaway: since deer had become extinct, these creatures evolved from the rabbits that humans had scattered so extravagantly around the world. So, there is the common rabbuck (*Ungulagus silvicultrix*) in temperate forests, the mountain rabbuck (*Ungulagus scandens*) as the smallest and most agile of the group, the arctic rabbuck (*Ungulagus hirsutus*), covered with thick fur—and so on. All these creatures—the falanx with its vicious teeth, the chirrit cutely nestling among branches, the chiselhead excavating burrow systems in tree-trunks with its massively developed incisors, and many more—were drawn to be functionally and ecologically *reasonable* descendants of whatever surviving species are left after humans.

Dixon only set a couple of scenes in the sea. In the polar oceans, the largest creature on the new earth is the vortex (*Balenornis vivipara*)—a whale-like and plankton-eating pseudo-baleen descended from the penguins, given the demise of whales. Almost as impressive is the 4-metre-long distarterops (*Scinderedens solungulus*), with paddle-like feet and long, forward-pointing incisor teeth used to pick shells off the sea floor; it is descended from rats.

It is a lovely, imaginative piece of work, with a serious subtext. After all, some kind of ecosystem has to arise from the ashes of the old world. Its component species would have to function under the same constraints as all life forms, since life began. But life—as biology and palaeontology tells us—can come up with an extraordinarily wide array of mechanisms and structures to allow organisms to make their living in the world. So, other than the vortex and distarterops—what might appear in the new oceans?

It is likely, as we have seen, that there will be a reef gap starting very soon. Reef gaps, in the geological past, have mostly lasted a few million years, before new reefs arose. Reefs tend to return in Earth history because, once formed, they are resilient, stable structures that enable very many creatures to coexist in a self-sustaining ecosystem. Reefs in the past have not all been coral-based. In the Palaeozoic there were reefs built by archaeocyathids and stromatoporoids (extinct organisms, both likely related to sponges). Reefs were also built, later in the Palaeozoic, by bizarre, tubular brachiopods: the richthofenids. These were good at what all successful reef-builders do—crowding out much of the competition for space and food and sunlight by massing together to make robust frameworks. Then, in the Mesozoic, molluscs followed the richthofenids by evolving the tubular rudist bivalves, and made in effect copycat reefs. Stony algae have made reefs, as have bryozoans (moss animals) too.

So what might make up the next generation of reefs, millions of years from now? Corals might bounce back, of course—the hexacorals that still build modern reefs have done so at least twice in the past, following the Cretaceous–Tertiary and Paleocene–Eocene crises. Perhaps the molluscs might come up with new designs: some descendant of the common mussel, perhaps, growing larger and moving offshore and binding tightly with its kin to convert areas of the sea floor into wave-resistant bulwarks. Or maybe the reefs of the future will be barnacle reefs, the evolution of these organisms having been given a helping hand by their struggles to cling to boats and resist the efforts of boat-owners to dislodge them. *Something* will eventually build reefs again, as long as oceans and complex life remain.

There will be many other niches to fill (and perhaps new niches to exploit) in the new oceans. If sharks continue to be relentlessly pursued, and most species eventually succumb (for contrary to their fearsome reputation, they are slow to grow and reproduce, and so now ecologically vulnerable) there will surely be a niche there—just as the dolphins are body doubles for the ichthyosaurs of the Jurassic. A radiation of shark-shaped bony fish seems most likely, but perhaps some mammal—a hyper-evolved sea otter, for instance—might step into the breach. One wonders, though, if something like the hammerhead shark will evolve again. It is having a thin time of it right now, and some creatures—as one can see in the fossil record—are essentially one-offs.

The Leaking Oceans

The oceans, as we saw in Chapter 3, have not simply been a constant mass of water sitting within the ocean basins that they rest in. The ocean basins are continually changing their shape, as sea floor is subducted down ocean trenches and created at mid-ocean ridges. The Earth is in effect cracked, and water can move through these cracks, into the depths of the Earth and back again. The evidence from the distant past of the Precambrian is that there are hints—and that is all there can be given the scrappiness of the rock evidence—that the oceans were more voluminous then.

So what is happening now? There have been several attempts to measure how much water is currently being forced down into the mantle along subduction zones (by looking, for instance, at the chemistry and water contents of the rocks along subduction zones), versus how much is coming back via the gas-charged volcanic eruptions that overlie these subduction zones and also much later (after a long journey through the mantle) at the mid-ocean ridges (by, say, measuring the amount of steam released during volcanic eruptions).

The amount of water going into the Earth seems to be something over a cubic kilometre of seawater each year (one estimate puts it at nearer 2 cubic kilometres), while the amount being erupted each year appears to be much less—perhaps something like a quarter of a cubic kilometre.[116] This suggests that something of the order of a cubic kilometre of water is being siphoned off from the ocean and is disappearing into the depths of the Earth, to simply dissolve into the minerals of the mantle.

The Earth's oceans hold some 1.3 billion cubic kilometres of water, so by projecting presently estimated rates of water loss into the future the Earth's surface will have lost most or all of its oceans a billion years from now.

It may not be quite as simple as that. The Earth's interior is changing through time. It is cooling slowly, and its chemistry is changing as volcanoes keep erupting magma, derived from that interior, on to the surface. Its chemistry and physical properties are changing, too, as water is mixed into it from the oceans. So the balance of water gained and lost between surface and exterior must change across long periods of geological time. Attempts to feed this more sophisticated understanding into a model suggest[117] that the rate of water uptake by the Earth's mantle will slow down over time—and that the Earth can never lose all of its oceans by that means, but will stabilize with over half of its volume still at the surface. A billion years from now, the oceans might only have lost a quarter of their volume by leakage into the mantle.

This would still look like a different Earth, and even this relatively small loss might have big implications for life on Earth. With a quarter of its volume gone, the oceans will not flood over on to the continental areas as they do today, but they will simply lap up against the relatively steep slope that separates ocean basin from continental mass. The continents will therefore loom larger, when viewed from outer space. That will mean the end of shallow seas, and so the end of many of the habitats that today we commonly think of as 'the sea', from the North and Irish Sea around Britain, to the Baltic, to the eastern seaboard of the US. On such an Earth, it will be deep water or nothing.

With less water, the sea will likely be more salty. Worse: many of the Earth's great stores of salts that were evaporated from the seawaters in 'saline giants' (Chapter 3) were stored on and around continental shelves. With these now standing high and dry and being eroded, their freight of salt will wash into the sea to make the ocean waters yet more briny. There will for sure be strong selective pressure for organisms to keep pace, and to adapt to yet saltier waters (think of the brine shrimps of the Great Salt Lake, for instance). Some will manage to do this, but many will not.

It seems that oceans may not disappear *into* the Earth, but they might be removed by other means.

A Hotter Sun and Thinner Blanket

Throughout its existence, the Sun has been very, very slowly becoming hotter. Three billion years ago, as 'the faint young sun', it gave off only about 80 per cent of the light and heat that it does today, and the Earth and its oceans kept from freezing by having a thicker atmospheric blanket of some combination of carbon dioxide, methane, aerosols, and perhaps simply a higher atmospheric pressure made up by more nitrogen.

In a billion years the Sun will be about 10 per cent hotter than it is today, as more of its hydrogen is consumed. The Earth, therefore, will inevitably become warmer than it is today. On that basis alone the Earth should, over at least the first part of the next billion years to come, become less prone to being glaciated and revert to conditions akin to those of the Mesozoic, with little or no polar ice, high sea levels, and flooded continents.

One consequence of warming, though, is that many chemical reactions at the Earth's surface will take place faster—and one of these is the combination of carbon dioxide in the atmosphere with rocks at the surface, to produce dissolved carbonate and bicarbonate ions that wash into the sea, ultimately to produce limestones. This could lead to atmospheric carbon dioxide falling to low levels. For some hundreds of millions of years, this effect might be sufficiently strong to overcome the increasing heat of the sun.

Entangled in this, though, is the effect on life: an important regulator of carbon dioxide levels on Earth, both through its processing and storing of carbon, and also—via the catalytic effects of surface microbial colonies and the presence of soils, plant roots, and such—increasing the rate of chemical weathering (and therefore of carbon drawdown) at the surface. If carbon dioxide levels fall sufficiently low then plants may not be able to photosynthesize and would die out, likely acting as a brake on the further drawdown of atmospheric carbon dioxide by surface weathering.

There will be other factors at work, not least the changing pattern of oceans and continents, to alter the way in which ocean currents flow. It is a complex multi-dimensional puzzle, and the fight between icehouse and greenhouse conditions on our planet may continue for some time. But then, there will inevitably come the end of oceans.

The Moist Greenhouse

At some stage in the future, no matter what happens to atmospheric carbon dioxide levels, the Sun's heat will take the upper hand. This is the beginning of the end of the Earth's oceans. As the Earth warms, towards mean temperatures that will rise into the forties and then into the fifties Celsius, more and more seawater will evaporate and the atmosphere—warmer too—will become increasingly charged with water vapour. We humans, who love temperate conditions, would find conditions unbearably humid were we able to visit this inescapable future state of our planet. If we could cope with such conditions though, then the dying Earth would be undoubtedly dramatic. The stifling heat, sky-high humidity, and roiling clouds would generate extraordinary firework displays of lightning strikes[118] and rolling thunder—truly a spectacle worthy of the *Götterdammerung*, a twilight of earthly gods and earthly powers.

Water vapour is a powerful greenhouse gas, and its build-up in the ever-hotter atmosphere will raise temperatures yet higher. The warming will affect upper levels of the atmosphere, including, crucially, the stratosphere. It is the stratosphere today that acts as a lid on the Earth's water (see Chapter 3). A highly effective cold trap, it prevents anything but minuscule amounts of water molecules from reaching the top of the stratosphere. If the stratosphere can be kept sufficiently cold, then the Earth could preserve its water supply for many billions of years.

However, in a warming stratosphere water vapour molecules can drift into its upper levels, where they become vulnerable to being split into separate hydrogen and oxygen ions by solar radiation. Once in that state they—in particular the small, light hydrogen ions—can be stripped out into space by the solar wind. This is the moist greenhouse state, as elaborated particularly by the planetary scientist James Kasting of Pennsylvania State University.[119] On such a pervasively warm, humid world, an Earth's supply of ocean water can be siphoned off into space in something of the order of a billion years.

Another factor, over that time, will be the weakening and subsequent loss of the Earth's magnetic field. The Earth's iron-nickel core was originally completely molten, before beginning to freeze into the solid inner core and the liquid outer core, currents within which generate the Earth's magnetic field. The solid core, though, is growing as the liquid iron solidifies (currently at something like 5,000 tonnes a second). A billion years from now the core will be entirely solid, and so unable to generate an effective protective magnetosphere. It will be all the easier, then, for the solar wind to remove the Earth's water.

Ocean Remnants

The end of the oceans is not likely to be simple. A once-watery planet does not become a uniformly dry, dusty planet overnight—not even *geologically* overnight—unless, that is, another kind of greenhouse comes into play, and we will come to that shortly.

The Earth, even in its hotter future, will still have areas that are relatively colder at the poles, even while the equatorial regions bake. Even as the water is siphoned off from the oceans, there is still at least a further ocean's worth of water deep underground, dissolved into the minerals of the mantle. This water will slowly be released to the surface as the Earth's volcanism, within its newly emerging pattern, continues. It will help preserve the Earth's ocean remnants a little longer.

These remnants will shelter what have been called swansong biospheres.[120] The evocative term is new, but the concept is not. The swansong biospheres will probably be microbial, as the increasingly harsh conditions will likely have put paid to all species of multicellular animals and plants some time before. They will be in the polar regions.[121] Buried underground, out of reach of the fierce rays of the Sun, there may be caves, or partially collapsed lava tubes. Any air that is relatively colder and denser may gather in such places. With sufficient cooling, the water vapour may condense into underground pools and puddles, to give some of the last patches of more or less pure water on Earth, a last gift of the oceans to life on Earth, sheltering some of the last microbial communities.

Elsewhere, there may be the final puddles left in the remnants of the oceans, highly saline and nestling within hollows in thick salt deposits. The microbial communities surviving in these last ocean patches would need to be hardy, capable of tolerating not just the high salinities, but also swings in pH from highly acid to very alkaline, high and variable temperatures, intense sunlight, and high UV radiation (in a low-oxygen future Earth, the protective ozone layer would no longer be present). These last ocean communities may resemble, perhaps, the microbial mats that form around black smokers—regions on the sea floor through which hot volcanic waters gush today—or like those in the crater lakes of the volcanoes in the high Atacama Desert of Chile.

Runaway Earth

James Kasting's models revealed a further state, if temperatures were to ratchet yet higher, such that the oceans began to boil. In this state, as water rapidly converts to steam the atmosphere fills with water vapour faster than it can be lost at the top of the stratosphere. This is the fearsome state known as a runaway greenhouse world. Its high-pressure steam atmosphere will become a super-efficient insulation blanket. Sunlight will get in, but very little heat will escape.

On this runaway world, temperatures could rise to something near 1,200 degrees Celsius, hot enough for rocks to melt.[122] A magma ocean would form at the Earth's surface. This is the kind of fate that Venus is thought to have suffered a long time ago (Chapter 9). Temperatures will only begin to drop once all the water had gone.

A post-runaway world of this kind would look very different to one in which only a moist greenhouse had operated. Its hard, fused surface would likely look like that following the cooling of one of the more fluidal volcanic lavas—smoothed, shiny, perhaps wrinkled or cast into rope-like shapes here and there, with smoothly sculpted hills separated by flat-floored valleys into which the molten rock had puddled. There would be no soil, no sand, no sediment of any kind, other than perhaps the barely recognizable traces of some of the largest boulders. Much of the landscape would be blackened, rather like obsidian, with here and there some paler patches—for instance where sand deserts with their winnowed masses of quartz grains had melted into glass. This would, of course, be quite emphatically a stone-dead world.

There would be little surface sign that the world had once been sedimentary, or had possessed the finest collection of rock strata, by far, in the solar system (and probably one of the best in any star system). Perhaps there might be hints of strata in the landscape still—imagine partially melting the rocky bluffs of something like the Grand Canyon, so that only some partly fused vestiges of those strata-formed rock ledges remained. Even so, this is equivalent to wiping out much of the history of a planet, as rock strata are the tape recorders upon which a planet's history is written, in the code slowly learned by every geologist, of different types of sedimentary layering and fossils and preserved chemical markers.

Much strata—and therefore much planetary history—would remain, of course, at depth, below the reach of the blast-furnace-like heat that had seared the surface. On today's Earth, buried strata can reach the surface as the landscape is acted upon by the wind and the weather and agents of erosion, and as tectonic movements raise and lower sections of crust in different parts of the world. But this is now an Earth bereft of a hydrological cycle, without rain, hail, or snow. There will be some kind of an atmosphere, and there will be winds, but with no sand particles left upon this blow-torched Earth there will be nothing with which the Earth can abrade its molten carapace, except perhaps the occasional meteorite. And as for tectonic movements, on an old, dry planet tectonics might soon begin to operate differently, as we shall see in the next section. Those uplifted crustal masses and down-warped grabens might be, literally, a thing of the past.

The ancient strata would still be reachable by drilling—if any visitors from space would think of trying to see what might lie below the surface of such an outwardly hostile surface. They would have to be space visitors with time on their hands, for this would not be like the present Earth, which is chemically and biologically dynamic enough to immediately attract any passing interstellar travellers with a half-decent set of sensors fitted to their spaceship. This inert, melted lump, orbiting in the glare of a dying star, would not even attract a passing glance.

By whichever type of greenhouse, the Earth will become a dry planet.

The Earth after Oceans

An ocean-free Earth will not simply be the world we now know, deprived of water. The deep, dry ocean basins will be covered in salt crusts. The ocean trenches will represent the last deep seas of all, made up of the dense brines the formerly abundant seawater had become, and being filled with salt deposits some kilometres thick. The now-exposed oceanic ridges will form land-based mountain chains. For a while, they may still be stretching apart and erupting floods of basalt. At the edge of the ocean basins, the continental slopes may still be traversed by the degraded remains of the steep-sided submarine canyons that used to carry dense floods and slurries of sediment-laden water to the ocean floor. In the mountain chains that fringe the continents, there may still, for a while, be volcanoes like those of Popocatepetl and Chimborazo today intermittently and violently erupting steam- and gas-laden masses of ash and pumice high into the atmosphere.

The familiar, long-lived tectonic activity of the Earth would likely not persist for long. As the oceans disappear, the subducting plates will no longer be lubricated by the seawater that they used to carry with them. They will find their passage into the mantle more difficult and friction-ridden. The upper regions of the mantle, too, deprived of the water that used to be continually stirred into them, will begin to become more viscous and less able to flow. The extraordinary mechanism of Earthly plate tectonics would, geologically soon, grind to a halt.

It would be one of the most fascinating episodes of our extraordinary planet. It would be a planet not quite in its death throes (although precious little if anything would be left of the Earth's biology by then), for it would still have a few billion years of existence ahead of it. Just how would it work? The machinery of plate tectonics is unlikely to simply just stop, smoothly and without fuss. Any time travellers incautious enough to venture into the time of the great change would likely see some extraordinary landscapes, as ocean plates were driven towards an obstacle—the ocean trenches—that they could no longer pass through, or some parts of ocean ridges stalled while others kept splitting apart.

The painful transition into an Earth without plate tectonics would probably be followed by an eerie calm for perhaps some tens of millions of years. The Earth would be adjusting to a new situation. For even though its inner fires would be waning it would still have a heat release problem, and with the halting of plate tectonics that would need another solution.

One of the new features of the Earth without oceans and with gridlocked plate tectonics will be a change in the pattern of volcanism. As we have seen, volcanism is not just an accidental by-product of plate tectonics. It is a planetary heat control mechanism; a means of releasing the large amounts of heat being produced within the Earth, as radioactive elements continue to decay within the Earth's interior. There will still be too much heat being produced, even late in the Earth's existence as the amounts of radioactivity diminish, to be simply lost by conduction to the surface. The standard means of planetary heat loss is to melt rock in the interior and allow that to escape to the outside, where it can release heat to the exterior and into an atmosphere or outer space.

When the plate tectonic machine has seized up, the heat, for a while, will simply build up, causing magma to accumulate and pool at depth underground. Then one can envisage a less active, more patchy reprise of the heat-pipe Earth of early Precambrian times (see Chapter 3), with magma channelled to the surface via more or less stationary vertical conduits, to release its heat via very large shield-like volcanoes that may, perhaps, reach the size of ancient Olympus Mons on Mars, or some of the volcanoes on Venus (see Plate 5).

The hot, dry, baked Earth—nothing at all like the beautiful blue planet of today—will be long-lived, for there will be something like 3 or 4 billion years yet of this lifeless existence. Then the Sun, having finally exhausted its reserves of hydrogen, will begin to puff out into its red giant state, out beyond the orbit of Venus. The Earth may then simply be engulfed, or perhaps it might be expelled into outer space by the final blowing off of the outer layers of the Sun, to drift forever in the dark.

This is the normal fate of what, for part of its existence, would surely have been one of the jewels of our galaxy; but it is the fate of many planets to bloom only briefly with all the possibilities that liquid oceans can offer. Even the other planets and moons of our own solar system, orbiting their unremarkable yellow star, can show the many forms and many histories that planetary oceans can have. It is these that we turn to next.

9

Oceans of the Solar System

Early Days

What kind of worlds, and oceans, lie beyond Earth? This question has fascinated people since they first began to gaze into the star-strewn heavens. Democritus (~460–370 BCE) was among the first to leave a record of his musings. A man who chose to smile on life rather than frown on it, he was a cosmic pluralist, and imagined many worlds beyond Earth that had grown from the atoms that whirled across the heavens. Those distant worlds grew, and decayed, and some were destroyed in collisions with others—while some, to him, were 'bare of animals and plants and all water'. Given what we know now, these were inspired intuitions. Nevertheless, they did not find favour with a younger and ultimately more influential man, Plato, nor with his even more famous student, Aristotle, both of whom considered that there was no world in the universe but our Earth. They knew of the planets, but thought them to be perfect bodies in the heavens.

This worldview of Plato and Aristotle was to govern—or perhaps to muzzle—scientific thinking for the next millennium and more. Then, with Galileo and the invention of the telescope, the spirit of evidence-based inquiry into the heavens began—although not without opposition—to stir once more. Galileo was born in Pisa in 1564, into a Western world where most people still believed the Sun, the Moon and everything else in the skies was circling the Earth at the very centre of creation, at the centre of the universe. Science was in its infancy, and it was not unusual for a university professor to be at once mathematician, astronomer, and astrologer.

Within a year of the Dutchman Hans Lippershey's invention of the telescope in 1608, Galileo had built his own rudimentary telescope and pointed it towards the heavens. Galileo observed the mountains of the Moon, the moons of Jupiter, and the crescent phases of Venus. His observations led him to support the heretical ideas of Copernicus, and so sealed the fate of the Sun-centred solar system. They also famously drew the attentions of the Inquisition—and Galileo lived out his final years in house imprisonment. But what were the planets like? The telescopes of the day gave little clue as to the nature of those distant worlds.

That did not stop speculation. The ferocious workaholic (typically being at his desk some 14 hours each day) George-Louis Leclerc, the Comte de Buffon (1707–1788), was arguably the first scientist to stretch time, as well as space, in invoking past geological epochs of an Earth for which he ascribed the then unthinkable age of 75,000 years—far beyond the few thousand years that could be derived from study of the Bible. As with Galileo's insights, it was a revolutionary step—and as such it also attracted the criticism of the religious establishment of the day. Buffon, though, avoided Galileo's fate. He had considerable influence, and considerable diplomacy too—something that was lacking in the more hot-tempered Galileo (privately, Buffon thought that the Earth might be as old as 3 million years, but he wisely kept this yet more outlandish figure to his personal notebooks).

Buffon considered the Earth—and the other planets too—to be cooling fast, and that they went through a similar history in which the planetary surface cooled, oceans formed, complex carbon-based life was generated quickly and spontaneously, and this, some thousands of years later (according to his published timescale), was doomed to perish as the planet froze. To Buffon the planets of the solar system were therefore broadly alike, and the kind of evolution of life and environment on Earth was nothing special, but was repeated in general outline among planets everywhere.

The Pioneer

A century after Buffon, the great Swedish chemist and polymath Svante Arrhenius (1859–1927) considered the question again—only by this time astronomical techniques had improved. Telescopes provided clearer pictures, and showed the existence of not just many more stars but of galaxies too. Somewhat confusingly, these were then called 'nebulae', while the dust clouds that we call nebulae today were glimpsed as 'empty spots' that, some thought, might have been due to 'opaque mist-formations' that blocked the starlight. Arrhenius, presciently, regarded these 'mist-clouds' as the breeding ground for new stars, and was aware too that the Milky Way was a spiral 'nebula' of enormous scale—his estimate of 100,000 light years being not far off present estimates. The universe, by his day, had hence become gigantic.

Furthermore, light from the stars and planets could be analysed to give clues as to the chemistry of distant objects. Arrhenius knew, therefore, of the abundance of hydrogen and helium in the universe, and deduced that Mercury, like the Moon, had essentially no atmosphere—and no water. With Mars he was well aware that it changed its appearance though time, and he knew of the large Martian polar ice caps first glimpsed as light patches by the Italian-born French astronomer Giovanni Cassini in 1666. A hundred years later, the British astronomer William Herschel (see Chapter 1) saw the southern ice cap shrinking and expanding with the seasons, suggesting the presence of ice. Arrhenius knew, too, of the debate surrounding the canals of Mars. The publication of Giovanni Schiaparelli's map of Mars in 1877 had depicted a complex network of natural channels or *canali* on the surface of the planet. This was a catalyst for some visionaries, principally the New England industrialist Percival Lowell, to see evidence of civilization on Mars.

Lowell was possessed of an eclectic range of interests, and after spending ten years in the Far East, where his cultural observations were recorded in a series of books, he returned to the United States to pursue his great love of astronomy. Lowell built a state-of-the-art observatory near Flagstaff in Arizona: the observatory is still operational today, although the original 24-inch refracting telescope he used is now something of a museum piece. Lowell dedicated the latter part of his life to observing the surface of Mars. He saw Schiaparelli's natural channels as canals, which he interpreted as a vast network of irrigation for a drying planet. He even thought that he saw cities connected by the channels, and patterns of colour change on the surface through the year that he interpreted to be changes in vegetation. The possibility of water and life on Mars fired the imaginations of nineteenth- and twentieth-century novelists who dreamed of fantastical, intelligent Martians, perhaps most famously depicted as the callous and deadly invaders of H. G. Wells's *The War of the Worlds*, armed with heat rays and tripod-shaped fighting machines.

Arrhenius dismantled Lowell's claims with forensic skill. For a start, the planet was simply too cold. The temperature at the surface could be calculated ('easily', he said—but then he was a prodigious mathematician) from a consideration of the solar heat reaching the surface of that distant and small planet, a little over half the diameter of Earth. He obtained a value of −37 degrees Celsius, which was far lower than Lowell's figure of +10 degrees Celsius. Lowell, though, had invoked a thick atmosphere, charged with heat-trapping gases such as water vapour.

Arrhenius, among his other accomplishments (he won a Nobel Prize in 1903 for his early work on the nature of electrolytes in solution[123]), was a pioneer in the study of greenhouse gases, being the first to propose that past changes in the Earth's carbon dioxide concentrations had caused the Ice Ages. He examined the available evidence for a thick, heat-retentive atmosphere on Mars by considering the most painstaking spectroscopic measurements of Mars then made to try to detect water vapour. From Earth this was difficult to measure, then as now, because of the obscuring effect of water vapour in the Earth's own atmosphere (see Chapter 1). Nevertheless, to Arrhenius the data were clear. The Martian atmosphere was thin, with very little water vapour—and so the planet was essentially deep-frozen, with little chance of any but the simplest and sparsest life forms. Arrhenius did not dispute that huge natural valleys might exist, but ascribed them to tectonic activity and not to an alien civilization. He also thought that at the base of these natural canyons, and in low-lying areas, there might be a little liquid water—especially if it was a sufficiently concentrated brine, charged with dissolved salts, to resist freezing. Regarding Lowell's vision—of cities 50 times bigger than London, and of a planet made red by autumnal leaf change, as in New England during the fall—he was politely scathing. The trouble with explanations like those, he said, is that they explain everything, and therefore they explain nothing.[124]

Arrhenius had more trouble with Venus. This planet is the brightest object in the night sky after the Moon. It had been known from ancient times: a seventh century BC Neo-Assyrian cuneiform text from the British Museum records astronomical observations of Venus, the 'bright queen', from 1,000 years earlier. At 95 per cent the diameter of Earth, and a little over 80 per cent of its mass, it is Venus and not Mars that is the true sister planet of Earth. Nevertheless, this planet, dedicated by the ancients to the goddess of love and sensuality, long retained her mystery beneath a thick concealing veil.

Even the best telescopes of Arrhenius's day showed just a dazzling, brilliantly white lustre. Arrhenius thought that the thick cloud cover was of water vapour. On this basis, he calculated that the planet was warm—about 47 degrees Celsius on average, and very humid ('about three times that of the Congo'). The planet, he said, was dripping wet, with dense rain clouds giving rise to violent rainfalls that poured down rapidly, eroding river valleys into the wide Venusian oceans. These conditions, he said, would support life, and he speculated on abundant but simple plant life that grew rapidly, and died and decayed equally rapidly, amid swamps that might have resembled the Carboniferous coal swamps of an ancient Earth.

This vision helped spawn, over much of the twentieth century, many Venusian science fiction epics where the breathless drama is set among jungles and seas. This was usually done quite shamelessly and with little concern for scientific veracity, as by Edgar Rice Burroughs (best known for the Tarzan books). His hero, Carson Napier, first wound up on Venus by accident after aiming for Mars (he had forgotten to allow for the Moon's gravity), and then proceeded, on that exotic water world, to do battle with pirates and zombies and the hideous Cloud People, all the while wooing the fair Venusian princess Duare. C. S. Lewis got in on the act with an ocean-covered Venus as backdrop for a clash between good and evil in *Perelandra*. Even such a scientifically literate science fiction writer as Isaac Asimov could set his 1950s epic *Lucky Starr and the Oceans of Venus* (memorable for a giant carnivorous orange patch that catches its prey using a jet of water) on a temperate, water-rich, habitable planet.

If you really want to find out what a place is like, though, you have to go there. Our understanding of the solar system has been transformed by the complex, ingenious, fragile space probes that, since the 1960s, have been penetrating deep into its mysteries: from Mercury as far as Neptune. As we write, the New Horizon satellite is heading for Pluto (formerly known as a planet) and will make contact in 2015—and then head for the Kuiper Belt beyond. These have been real voyages of discovery, and have opened up worlds that are stranger and more diverse—and in places far more ferocious—than the imagined planets of the science fiction writers. Some have oceans perhaps larger than Earth's. Some had oceans in the deep geological past. Some might yet acquire them. We can explore them now, moving from the regions close to the Sun into the chilly outer fringes of our own star system.

The Space Exploration Era

Mercury, at its closest, orbits just 46 million kilometres from the Sun (28.5 million miles), and lies far from the solar system's 'snow line' (see Chapter 1). It is a rocky world of extremes, cratered like our Moon, with temperatures swinging from a furnace-hot 425 degrees Celsius during the day to nearly minus 150 degrees Celsius in its planetary night. It certainly does not support anything that we might call an ocean, no matter how hard we stretch the definition of that word. Nevertheless, in the floors of craters near Mercury's north pole, in patches that are permanently shaded from the Sun's heat, scientists at Caltech and NASA's Jet Propulsion Laboratory in the early 1990s used radar to identify bright spots that might be ice—and, if so, perhaps relics left from the impact of ancient comets. Recently, these bright patches have been confirmed by detailed radar images from the *MESSENGER* spacecraft, which has been in orbit around Mercury since 2008. It is still not certain, though, that the bright spots do indeed represent water ice (they might be of another volatile substance, such as sulphur). If it is water, it must—even though permanently shaded—be covered by a thin layer of rocky debris to prevent it from sublimating into space.[125]

Mercury has never supported water oceans. Its proximity to the sun's solar wind, its extreme temperatures, and its low gravity means that gases in its thin atmosphere easily escape to space. To find traces of oceans on other worlds in our solar system we must visit our other rocky neighbours, Venus and Mars, and delve into their distant past.

The Bright Queen of Hell

The earliest satellites aimed at Venus had little luck. The Russian *Venera* spacecraft probably got close in 1961, within 100,000 kilometres or so—but radio contact had been lost soon after it left Earth. The American *Mariner I* did not even get that far, blowing up on the launch pad, but in 1962 *Mariner* 2 got close enough for its microwave and infrared detectors to pierce the cloud cover to measure the surface temperature as higher than 400 degrees Celsius. At that moment, the dream of a habitable sister planet died in human minds—as did the (respectable) supply of at least one subgenre of science fiction. Such temperatures are far beyond the limits of tolerance of even the most heat-loving bacteria on Earth.

Venera 3 crash-landed on Venus in 1966, the first human-made object to fall upon that planet, but its communication system failed before it could return any information to Earth. The next three *Venera* spacecraft got a little farther, and revealed the reason for the extreme surface heat levels: the extraordinarily high pressure of the carbon dioxide atmosphere, equivalent to 90 Earth atmospheres. The atmosphere of Venus is laced with sulphuric acid for good measure—it is droplets of this corrosive substance that make up the thick cloud cover around the planet. The *Venera* craft were sturdily built; to one visitor to the Soviet workshops, they brought Henry VIII's armour to mind. Yet they were crushed by that atmosphere before they reached the surface. The first images of the surface of Venus were finally garnered by the specially strengthened *Venera 8, 9,* and 10 spacecraft, which functioned for up to an hour after landing, and sent back grainy black and white pictures of a barren rocky landscape.

Spacecraft continued to probe the physical and chemical characteristics of this fantastical and hostile planet. The *Magellan* missions of the 1990s used radar to map its topography in exquisite detail, revealing a landscape of volcanoes, some bizarrely shaped like pancakes, and lava channels thousands of kilometres long (in that furnace heat magma takes a long time to congeal, even when at the surface). As for water, there is almost none: merely the barest traces as water vapour in the atmosphere. The surface is, mysteriously, only lightly cratered, leading to suggestions that every half-billion years or so Venus 'resurfaces' itself, almost literally turning itself out in a welter of outpouring magma. That makes some kind of sense, for a world without water cannot have plate tectonics (see Chapter 2). Plate tectonics is, at heart, a gentle, steady, planetary heat release mechanism which we on water-rich Earth are lucky to have. Venus, therefore, seems to have evolved another, far more terrifying, form of intermittent heat release (as if its 'normal' conditions were not bad enough).

Venus seems not always to have been this torrid. Given that it is much like Earth in its composition and mass, the processes that produced water on Earth should also have produced water on Venus: from volcanoes and incoming comets from space. How then did Venus end up with only about a thousandth of 1 per cent of the water within the Earth's atmosphere and oceans?

There is evidence that water was once present on Venus, and was perhaps even plentiful—and this planet might once have been akin to the imagined world of Svante Arrhenius (if not that of Edgar Rice Burroughs). This former hydrosphere is suggested by the unusual chemistry of Venus's atmosphere. Of the small amount of hydrogen bound to the tiny content of water in the atmosphere, there is a strikingly high abundance of the heavy isotope deuterium: about one in every 10,000 hydrogen atoms (which means that one in every 5,000 water molecules contains a deuterium atom and is thus 'heavy water'). This ratio may seem small, but it is nevertheless 100 times greater than on Earth.[126]

The discovery of the deuterium enrichment was quirky in the extreme. When the NASA *Pioneer* mission parachuted down in 1978 it had a mass spectrometer on board to analyse the chemistry of the Venusian atmosphere. A drop of sulphuric acid from the clouds condensed in the inlet to this instrument, seriously compromising virtually all of the data. However, the spectrometer then analysed the acid droplet as it evaporated—and discovered the very high deuterium values.[127]

Assuming that the original ratio of deuterium to hydrogen was similar on Earth and Venus—and this is a reasonable assumption, as their waters probably came from much the same sources—Venus has lost at least 99.5 per cent of its original water.[128] This has simply leaked into outer space and been carried away by the solar wind (with more of the deuterium having been retained because, being heavier, it is less easily lost from Venus's atmosphere). This may have been the greatest planetary tragedy (to date) in the history of the solar system. For, while the fate of Mars would have been inevitable in any case—as Arrhenius realized—perhaps Venus, under slightly different circumstances, might have clung on to its seas (and life?) for longer. How then did Venus dehydrate and die?

When Venus was young the Sun was fainter than it is today—faint enough for Venus to retain something resembling an Earth-like supply of water. If so, seas or oceans must have existed on its surface. Perhaps, too, if stores of that planetary lubricant, water, were plentiful enough, there may have been some form of plate tectonics, and therefore ocean basins too.

As time passed, the young Sun converted more hydrogen to helium and shone more brightly. The oceans of Venus warmed, and yielded more water vapour to the atmosphere. This greenhouse gas, acting in concert with carbon dioxide and probably other gases such as methane, warmed the surface even further—and this drove ever more evaporation from the surface of Venus. The water molecules rose high into its atmosphere, into regions where ultraviolet radiation could split them into their component atoms, to be carried off by the solar wind.

As the water supplies grew smaller, and the Venusian oceans shrank, the opportunities to remove carbon dioxide from the atmosphere by rainfall and then converting it into Venusian limestone or Venusian coal reduced, so the level of that greenhouse gas rose too. At some point, the vicious circle turned past the point of no return. The water disappeared, and levels of carbon dioxide kept rising, released by volcanoes or by the thermal destruction of any carbon-bearing rocks that had managed to form. Today, that process has reached completion. The amounts of carbon in the atmosphere of Venus now are roughly equivalent to that of the carbon stored underground on Earth as limestone and coal.

Venus today lacks a strong, protective magnetic field and therefore easily loses any remaining water to the solar wind. Therefore, was the breakdown of the magnetosphere a key step in the loss of Venusian oceans? It is likely that Venus once had a magnetosphere like Earth's, as it accreted from similar materials to Earth and probably possesses an iron core. For part of its history, this iron core likely generated a strong magnetic field, and for the first billion years or so of Venus's history the atmosphere could have been protected from the ionizing radiation of the solar wind. At some point the core of Venus—perhaps because of the planet's lower mass—stopped functioning. Then the water in the Venusian atmosphere simply leaked away.

A World of Rust

The first spacecraft images of Mars, taken as *Mariner 4* flew past the planet in 1965, seem blurred and uninformative to us today. Yet they were enough to decisively prove Arrhenius right, and Lowell wrong. The pictures showed an essentially dead, dry, cratered, and ancient surface. It did not show that life is *completely* absent (that question is still open), but it dismissed ideas of alien forests let alone alien civilizations—or present-day alien oceans. Those first images have, over the succeeding decades, been superseded by ever more detailed ones, which show lovely, complex detail of the surface. Even better ones are arriving as the new lander, *Curiosity*, having survived the 'seven minutes of terror' of its elaborate landing procedure, has turned its high-resolution cameras on its surroundings. Yet these images—even those taken of rock strata from centimetres away, by cameras mounted on landing craft—have left the role of water in Mars's geological past still tantalizingly unclear.

There is no doubt that Mars does possess some water, which makes up the bulk of its icecaps (and these also have a surface layer of frozen carbon dioxide that sublimates seasonally into the atmosphere and then refreezes). There is enough water in those icecaps, if melted and somehow kept liquid (something that does not happen easily because of the very low atmospheric pressure, about 1 per cent of Earth's, and mostly carbon dioxide at that), to cover the Martian surface in a shallow sea a little over 10 metres deep. There is also significant water in the rocks at depth, as we shall see, frozen as permafrost and buried ice masses. There may be liquid aquifers in the warmer, more highly pressured environment deeper down (although none have been detected yet); if so, this is the most likely habitat for any simple life forms that may exist today on Mars.

There is also no doubt that a lot of water has flowed across the surface, in the geological past, in the first billion years or so of Mars's history (see Plate 6). The evidence for that has become ever clearer as the images from the successive space missions have accumulated (three are currently in orbit: the *Mars Reconnaissance Orbiter*, the *Mars Express*, and the *Mars Odyssey*—the last of these already more than a decade in harness and still sending back data). There is now a cornucopia of different kinds of imagery: photographic, infrared, radar, gamma ray, and more. Much of it is freely available, so anyone with a computer and Internet connection can fly virtually across the surface of Mars and gaze down. What does one see?

The evidence of a more dynamic past is clear. The planet has a great divide. Its southern hemisphere is dominated by rugged highlands—within which the greatest mountain in all the solar system, Olympus Mons, rises to over 23,700 metres. By contrast, the north forms a smooth low-lying plain, much of which is extraordinarily flat. This contrast, referred to as the 'Martian dichotomy', is probably very old. The lowlands may have formed by the titanic impact of a Pluto-sized object more than 4 billion years ago, during the violent early history of our solar system.[129] The border zone between these provinces, and the adjacent parts of the highlands, are traversed by canyon- and valley-like structures, large and small, which grade down towards the northern plain. In places, these valleys end in delta-like structures, the surfaces of which have small distributory channels on them.

They look like the dried-up remnants of rivers flowing down to a sea. And, indeed, the northern lowlands have been interpreted as the remains of a substantial northern ocean, christened the Oceanus Borealis. Such an ocean, in the first billion years of the history of Mars, could have driven a hydrological cycle that created the rain that fell on to the adjacent land, sweeping in eroded sediment and recycling the water, for the cycle to begin anew.[130]

Did such an ancient Martian ocean really exist? Or is it simply a fanciful interpretation, akin to Percival Lowell's visions of Martian canals? As the pictures of this enormous region grew ever more detailed, not all of the evidence has pointed towards a former ocean.[131] The wide northern plains are not everywhere covered in fine sediment, as one might expect on an ocean floor far from land; large boulders lie scattered about, here and there, and in some places there are clusters of ridges and shallow depressions, while elsewhere there are regions of jumbled and collapsed terrain. Further, at the margins of the presumed ocean, there does not seem to be much in the way of well-defined cliffs and stranded beaches. Some people therefore thought that the northern plains had never been submerged under deep water, but rather represent volcanic plains of ash and lava, while another idea was that these flat areas had been 'resurfaced'—that is, smoothed—by long-lived permafrost activity.

More recently, evidence in favour of a former Martian ocean has once more turned up. A number of the canyon-fed deltas at the edge of the southern highlands occur at close to the same elevation, suggesting that—for a short time at least—this marked the position of a shoreline.[132] The boulders scattered across the lowlands may have been brought in by icebergs,[133] since any sea on Mars, so far from the faint early Sun, would likely have been a frigid one, with a shifting and fragmenting carapace of ice. Some of the linear depressions resemble the kind of grooves that large icebergs can gouge on a sea floor.[134]

The areas of collapsed terrain also look strikingly similar to sonar pictures taken of parts of Earth's deep-sea floor, where gigantic slabs of waterlogged strata have slid down gentle slopes, breaking up as they did so.[135] The balance of evidence seems to be tilting back in the direction of some kind of former ocean, however short-lived. Most

FIG. 18. Formed perhaps 3.8 to 3.5 billion years ago, the Gale Crater on Mars contains evidence of ancient lake deposits. In the foreground, the arm of the *Curiosity* rover is 2 metres long when fully extended. In the distance is the 5.5-kilometre-high summit of Aeolis Mons, which rises in the centre of the crater.

recently, the *Curiosity* rover (Fig. 18), on its way to climb the enigmatic, 5-kilometre-high Aeolis Mons in the middle of Gale Crater, has discovered the certain remains—the dried-up floor—of an ancient lake, termed Yellowknife Lake, that seems to have contained water that was not too markedly saline and that contained elements necessary for life, such as carbon, sulphur, nitrogen, and phosphorus.[136] So there was water for sure, ranging from lakes to possible oceans. Where did it come from?

On Earth today, volcanic eruptions release considerable amounts of steam because the magma itself can contain significant amounts of water—up to a few per cent—that is released when this molten rock nears the surface. Could the enormous Martian volcanoes, therefore, have been a source of water for the ancient ocean that seems once to have been on that planet? Clues here may be found in the very few meteorites found on Earth that show, by their chemistry, that they must have come from Mars,[137] having been blasted off the Martian surface by meteorites, drifted across space, and then fallen to Earth. These meteorites are mostly basaltic, and now contain very little water, but it had been suggested they had originally contained as much water as do terrestrial basalts, but that their history of shock and flash-heating had dehydrated them. Further study, though, indicated that they were *originally* poor in water, but that instead they contain a good deal of chlorine as a volatile phase—more than twice as much as in equivalent rocks on Earth.[138] Volcanism, it seemed from this, did not provide much of Mars's early water, which must therefore have originally come from the icy comets that crash-landed on that planet early in its history.

Yet these same meteorites have recently been re-examined, and the new studies focused on tiny crystals of apatite within them. These are relatively strong, leak-proof crystals that formed early in these rocks' history as they began to crystallize deep in the Martian magma chamber. The composition of the apatites suggests that the magma was relatively water-rich. The degassing then likely took place not in impact-shock on hitting the Earth, but much earlier, as the original magmas were ascending to the Martian surface.[139] So volcanic eruptions may have provided substantial amounts of water after all. (We suspect this debate will rumble on for some time yet, as the detective work continues.)

How long did that (presumed) Martian ocean last? Could the surface of a youthful Mars have been warm and wet, at least periodically, with a long-lived ocean and a hydrological cycle something like that of Earth? This could only have taken place below a dense greenhouse gas atmosphere, which could balance the reduced light and warmth of a faint young Sun. It is difficult to see how Mars could have made—and retained—such an atmospheric blanket. The evidence on the ground does not support such an idea, either. One symptom of long-lived water is the alteration of minerals to clay, both at the bottom of a sea and on land (where erosion soon carries that clay into the sea). Clays on the Martian surface have been identified and mapped by the orbiting spacecraft from their particular spectral pattern—but they are rare on the northern plains (although somewhat commoner in the highlands). This does not chime with a long-lived Martian sea floor that was being supplied with clay by long-lived river systems. In a cold and wet Mars though, where a cold ocean is fringed by ice, clay production and transport would be inhibited.[140] That ocean, and its attendant hydrological cycle, may also have been short-lived or intermittent. The erosional valleys of Mars are impressive, but there are few signs that those rivers were long-lived. A mature river system develops particular characteristics such as meander belts, and while there is one beautifully preserved—indeed iconic—ancient river meander on Mars, such structures are rare in general.

Those widespread early valley systems could rather be the product of brief, more violent, events. If most of the water on Mars has always been in the form of permafrost (perhaps with large aquifers beneath), this could have been a source for floods of water, following the largest of the many impact and volcanic events of this early era. Torrential outflow of such subsurface water may have created collapse structures and undercuts where the water exited the ground, to help shape the canyons and chaotic ground that make up a large part of those dramatic and puzzling present-day Martian landscapes.

Of these signs of cataclysmic outbursts of water, the greatest are concentrated in the Chryse Trough on the eastern margin of the Tharsis Bulge. The Tharsis Bulge is reminiscent of the giant flood basalt terrains on Earth, such as the Deccan Traps of India, some 65 million years old, and the even larger Siberian Traps that are some 250 million years old. The basalt lavas in these terrestrial examples erupted so quickly that both have been implicated in mass extinction events. If the Tharsis Bulge formed in a similar manner, the rapid eruption of huge quantities of volcanic lava at the Martian surface may have released large amounts of water.[141] The lava would have caused local heating of the crust, melting permafrost and so releasing yet more water to the surface. An ocean perhaps several hundred metres deep (or more: some estimates go as high as 1,700 metres) could then have rapidly formed on the northern plains.

Such an outpouring of water might have shifted the Martian climate, albeit briefly—from a cold, dry desert to a warmer, wetter state—restoring an Earth-like hydrological cycle with rainwater, rivers, deltas, and seas. The surface water would have released water vapour and carbon dioxide into the atmosphere—both greenhouse gases. As the Martian climate warmed, positive feedback mechanisms would have kicked in, when carbon sources long held within the Martian regolith and ice were released by melting. Thus, the Martian climate may, for a time, have warmed even more.

All too soon, cold would have returned. The increased moisture in the atmosphere would give rise to more snow at high latitudes: the icecaps would grow in size, and reflect more of the Sun's heat and light to cool the climate. Any increased rainfall, too, would have 'scrubbed' carbon dioxide from the atmosphere and allowed it to react with rocks, in silicate weathering, to cool the climate still further. A cold desert returned, perhaps to be punctuated again by subsequent ocean events, until Mars's internal engine cooled and volcanic activity dissipated, and the Oceanus Borealis finally shrank and died.

What were the waters of that short-lived ocean—or successive, temporary oceans—like? They were certainly salty—probably salty enough to act as a powerful antifreeze, remaining liquid at temperatures well below zero. But the brines were not like those of the Earth's oceans, because the dried-out salt deposits that remain include only tiny amounts of carbonates. There is nothing on Mars's surface resembling a limestone, despite the dominance of carbon dioxide in the thin atmosphere of Mars today. But there are large deposits of iron oxides and sulphates, a little like those that form in the strongly acidic water which oozes today out of coal and metal mines. Hence the oceans of Mars, it seems, over time became charged with sulphuric acid, and in such acid conditions precipitation of limestone would have been inhibited. Therefore, much of the carbon dioxide on Mars, instead of being converted into limestone rock as on Earth, simply drifted to the top of the atmosphere and was lost to space.

So this was an environment that seems to have been both physically and chemically harsh. These short-lived, acid, salty oceans would not have been ideal conditions for life—or at least for Earth-like life. However, we know that some types of microbes can survive and adapt to many different kinds of chemically extreme conditions. It remains to be seen whether fossilized microbes, or even living ones, can be found among the formerly water-rich provinces of Mars.

Where did the Mars ocean go to? In one interpretation it is still there, frozen under a thin layer of sediment.[142] Radar soundings suggest that layers of low-density material (such as ground ice) are present beneath the northern plains (while any aquifers of liquid water, if present, lie at deeper levels than those yet radar-sounded by the orbiting satellites). When the ocean froze, though, some of the water would have slowly sublimated—turned into vapour—and later would have settled on the polar icecaps. Some of Mars's original store of water would undoubtedly have drifted high into the thin atmosphere and then been 'sputtered' away by the solar wind, to be carried off into space.[143] Mars, after all, is considerably smaller than Earth, and so can lock in its water supplies less effectively. Finally, its magnetic dynamo switched off sometime in the first billion years, and so a protective magnetosphere would have been lost.

Never the setting for a balmy tropical paradise, Mars's oceans will not, perhaps, return until the dying days of the solar system. Then, the Sun will give out a final burst of heat as it swells into a red giant star, billions of years in the future. The ice caps and the remaining permafrost will melt and (briefly, once more) give rise to the last northern ocean of that planet. Soon after, that too will be blown into space.

Galileo's Moons

Far beyond the orbit of Mars, the thinly spaced rock fragments of the asteroids—in total amounting to only a few per cent of the Moon's mass—circle the Sun. Long thought to be dry (they are just the wrong side of the Sun's 'snow line'), ice has been detected even among these. Perhaps one day such ice masses will provide a crucial water resource for space-hopping metal miners of the future.

Far beyond the asteroid belt is the first of the gas giants, Jupiter, some 778 million kilometres from the Sun. It is enormous, having a mass more than twice that of the rest of the solar system (but only one-thousandth that of the Sun). Here we are well beyond the 'snow line'. Water has been detected in Jupiter's surface clouds, but only as trace vapour within the hydrogen–helium atmosphere. Deep down, there may well be more water (see Chapter 1), and this is one of the things that the NASA spacecraft *JUNO* will be looking for when it has a rendezvous with that planet in 2016. The question of water, though, is rather clearer with Jupiter's moons.

These moons were discovered by Galileo in 1610, and the realization that Jupiter was orbited by four worlds dealt a final blow to the Ptolemaic view of an Earth-centred universe. It was Galileo's rival, Simon Marius, who christened the four moons after illicit lovers of Zeus—Io, Ganymede, Callisto, and Europa. More Jovian moons have been discovered since: there are now 66 in total, although most are small with distant, eccentric orbits, probably being captured pieces of space debris. The Galilean ones, though, are substantial, with Ganymede being a little larger than Mercury.

For three and a half centuries after Galileo, their nature remained mysterious. From the 1950s on, analysis of their faint reflected light led to suggestions that some had a cover of water ice. In 1971 the astronomer John Lewis went further, and proposed that they might have sufficient inner heat sources (from radioactive decay in their rocky cores) to produce a liquid ocean beneath icy carapaces.

A couple of years later, in 1973, the first of the exploring spacecraft, *Pioneer 10*, flew past Jupiter and began to send images of that planet and its moons. More spacecraft followed—the *Voyager, Ulysses, Cassini*, and *Galileo* missions. They ignited a revolution in understanding that continues apace today. The sheer diversity of the moons of Jupiter, and of those of more distant planets, came as a tremendous—and exhilarating—surprise. Each one seems to have its own, quite distinct, identity. And among them there are true oceans.

Appropriately enough, it was NASA's *Galileo* spacecraft, which took off into space on 18 October 1989, that was to gather some of the key data. It was to fly over 4.5 billion kilometres before plunging into the atmosphere of Jupiter on 21 September 2003. On its journey it was faithful to its namesake, discovering new oceans beneath the icy shells of Ganymede, Callisto, and Europa.[144] Only Io, the innermost of the Galilean moons, has no water; any water would long ago have evaporated into space due to the moon's proximity to a hot young Jupiter. Instead, it is the most volcanically active body in the solar system, its interior being continually churned and melted by the powerful tidal forces exerted by its parent planet.

Of the watery Jovian moons, Europa is in many ways the most curious—and the most promising for those seeking life in alien oceans. Looked at as a whole, its surface is smooth, bright, icy—and geologically new, for few meteorite or comet craters mar its overall evenness. It formed—or was renewed—only some 50 to 100 million years ago, and therefore dates from the time of Earth's dinosaurs. Looked at more closely, it is criss-crossed by networks of dark lines with, here and there, patches of jumbled 'chaotic terrain' that look like masses of angular icebergs encased within sea ice (see Plate 7). The surface is clearly a dynamic one[145] which has undergone a long history of fracturing and healing. Europa's patchwork ice cover is able to flex and shatter in this particular fashion because beneath it lies an ocean estimated to be of the order of 100 kilometres deep, detected by magnetic patterns—which gives it roughly twice the bulk of the Earth's oceans. And beneath that is the main part of the moon, which is rocky with an iron core.

Powering the whole thing, as with Io, is the energy generated by the tides raised by Jupiter. This will be driving some kind of geological activity in the rocky heart of the planet. The tidal forces are thought to generate long low waves, termed Rossby waves, that propagate through the ocean of Europa, the kinetic energy produced being enough to keep the oceans liquid—and perhaps even lukewarm (although, with a surface temperature below −160 degrees Celsius, the external ice is rock-hard). The flexing produced as the moon is stretched and squeezed generates the fractures, the pattern of which shows that the entire icy carapace, with no attachment anywhere to the rock far below, is literally slipping around the moon's rocky interior, making a full revolution about every 12,000 years. The chaotic areas seem to be due to eruptions of warm, softer ice from below, perhaps associated with lakes of water entirely enclosed in the icy carapace. The water is salty, and the fracture lines seem to be enriched in crystallized salts—perhaps magnesium sulphate (Epsom salts) and sulphur compounds, to give them their dark reddish colour.

Among Earthbound scientists Europa has generated fierce debate. For instance, one question is whether it has a 'thick' surface ice carapace of perhaps 30 kilometres—which would mean that the water below never reached the surface—or a 'thin' one of only a few kilometres, where water could now and then break through, perhaps to form the 'chaotic terrain'.

There is, of course, the hot question of whether life might exist on Europa. Of water, there is no shortage. There may be more or less steady supplies of energy, too, although not the Sun's energy (so far out, and beneath thick ice, this is a world of pitch darkness). Any supply would be of chemical energy generated far down on the ocean floor, perhaps akin to the black smokers on the Earth's mid-ocean ridges, around which entire chemosynthetic communities cluster. There may, surprisingly, be free oxygen—for that molecule makes up most of Europa's tenuous atmosphere, formed from the splitting of water by radiation. Over time, sufficient amounts of this oxygen may have been carried down into that deep ocean to allow aerobic metabolism in any organisms that might live there.

It will be a long time before we find out. There are plans for further spacecraft missions, although it will be years before they arrive. There are plenty of ideas (it is the funding that is the problem). One marvellously creative idea that sadly has not gone much beyond the drawing board is the sending out of a rocket ship, heated by nuclear power, to melt through the ice (as a 'cryobot') and emerge as a submarine (or 'hydrobot') in the waters below. Today's science fiction, hopefully, will become science fact for our grandchildren.

It will be harder, on Jupiter's other Galilean moons Ganymede and Callisto, to reach through the icy crust down to the liquid water oceans that lie beneath. Ganymede, despite being the largest moon in the solar system, with a liquid iron core, lies farther out from Jupiter and so has a lower energy input from the tidal forces. The ancient icy crust (making up around half of this moon's mass) is up to 1,000 kilometres thick. Part of it is highly cratered and truly ancient—perhaps up to 4 billion years old—while part is younger, grooved and striated like Europa (although it is more cratered, and hence older); this doubtless reflects some phase of tectonics in Ganymede's deep history. Somewhere within this icy crust, hundreds of kilometres down, lies a layer of water that is salty enough for its electrical conductivity to have been detected by the *Galileo* spacecraft.

Callisto is far enough distant from Jupiter to have negligible tidal heating. Its icy crust is also thick, and is everywhere a mass of craters that likely date to near the beginning of the solar system. It seems never to have been heated enough for its rocky parts to differentiate properly into core and mantle, nor has there been anything like the surface tectonics shown by its sister moons. Nevertheless, even here there are indications of an ocean, over 100 kilometres down, kept liquid by the small amounts of radioactive heat generated by the rocky material lying deeper still.

The Oceans of Saturn's Moons

Stretching over 280,000 kilometres of space and up to a kilometre thick, the rings of Saturn are made mostly of water ice, some fragments being as big as terrestrial icebergs. Hypotheses about the origin of Saturn's rings range from the demise of an ancient moon or moons blown apart at the time of the Late Heavy Bombardment, 4 billion years ago, to a much more recent origin. Like those of Jupiter, most of Saturn's 62 moons are rich in water, although in the form of ice and not as oceans. But there are exceptions to this, even in these distant regions. The NASA spacecraft *Cassini* has shown that active, ocean-bearing planetary bodies can exist even 1,430,000 million kilometres or so from the warmth of the Sun.

To have Giovanni Domenico Cassini's name attached to this mission was apt—unavoidable, really, given that he discovered four of Saturn's moons, although really this was just a grace note in the life of one of Europe's most outrageously talented and effective scientists. His wit and social skills rivalled his scientific achievements. At the age of 25, in 1650, Cassini was made professor of astronomy at the University of Bologna, Italy, by Pope Alexander VII and, to the chagrin of his elder colleagues, was immediately paid the highest salary there. His talents were sought to attend to delicate matters of state as well as those of science—for instance to act as escort to Sweden's Queen Christina on her visit to Rome when she wanted to convert to Catholicism. A brief romance seemingly flowered between the two on the journey, to add to the storybook brio of Cassini's life. Invited to France to advise on Paris's new observatory, he immediately became a favourite of Louis XIV, the Sun King, and stayed there. Unperturbed (and unmoved by further grumbles from Cassini's Italian colleagues), the Pope carried on paying his Bologna salary. In his life, Cassini unravelled the structure of snowflakes, found that the speed of light was finite, discovered a gap in Saturn's rings (and as we have seen, discovered Mars's icecaps), and measured the size of France, ingeniously exploiting Galileo's discovery of Jupiter's moons as a measure. The last of these did not overly please the Sun King, as France turned out to be about a fifth narrower than previously thought. Poor recompense, he said, for treating his astronomers so well.

On Christmas Day 2004, the spacecraft *Cassini* released the European Space Agency's *Huygens* space probe[146] on its journey to the surface of Titan, the largest of Saturn's moons. *Huygens* entered the atmosphere of Titan on 14 January 2005. The saucer-shaped probe gently descended with the help of a parachute through a dense Titan atmosphere composed of nitrogen, methane, and small amounts of argon—this atmosphere is so cold that it contains no water vapour. Indeed, Titan's atmosphere is locked in a primitive state, with a composition perhaps similar to that of the early Earth. On Earth, though, where water vapour is active, carbon compounds in the atmosphere are rapidly oxidized to carbon dioxide. Not so on Titan. At mid-altitude in Titan's atmosphere (above 200 kilometres from the surface of the moon) *Huygens* encountered a thick smog where photochemical reactions produce an organic rain of methane together with nitrogen-containing aerosols. These fall steadily on to the surface of the moon, creating a strange landscape of rivers and lakes, carving valleys between mountains of water ice that are strangely reminiscent of rocky valleys on the surface of Earth.

As the rain falls through Titan's atmosphere it must create rainbows through the prism effect of methane droplets, which are as transparent as the rain on Earth. These rainbows will rarely be within the spectrum that humans see, as it is difficult for light to penetrate Titan's thick, hazy atmosphere. Future explorers from Earth might take with them night vision glasses, to see rainbows in the infrared spectrum—for radiation at this wavelength penetrates Titan's atmosphere more easily. The spacecraft *Huygens* itself saw the Titan landscapes in infrared, to help penetrate the haze.

After 2 hours and 27 minutes Huygens completed its descent and settled on to the surface of Titan, encountering a damp sandy surface—the damp sand here comprising grains of water ice moistened by liquid methane—and being rocked by gentle winds. For 90 minutes, peering through that thin haze, in the dim light (to an astronaut, it would be akin to a moonlit night) it sent back pictures of eroded, stream-transported ice pebbles stretching into the distance, and information on the atmosphere. It remains the most distant landing craft ever sent by humans.

Titan is the second largest moon in the solar system after Ganymede. It may have a complex, multi-layered water ice skin and a rocky inner core. Its skin includes a layer of water and ammonia that remains liquid at temperatures as low as −97 degrees Celsius. This deep-lying fluid is perhaps the source of much of the methane in the thick atmosphere of Titan, an atmosphere that is uniquely dense for a planetary moon at a little short of 1.5 times the atmosphere of Earth. Titan rotates slowly, in tandem with its circuit around Saturn of 15 days and 22 hours. But its atmosphere is rotating much more rapidly. Despite the gentle breezes of its surface, at an altitude of 120 kilometres the winds in the atmosphere are speeding along at more than 400 kilometres per hour. Titan is ferociously cold too. Its surface temperature reaches down to −179 degrees Celsius, much colder than the coldest place on Earth and cold enough to allow its methane seas.[147]

These hydrocarbons, falling as rain on the surface and being channelled through streams and rivers, accumulate as lakes and small inland seas: Titan is the only place in the solar system, other than Earth, where liquid persists at the surface (see Plate 8). This liquid is made of mixtures of ethane, propane, methane, and butane mixed with nitrogen, all these being stable under the ambient atmosphere of Titan. Of these seas, the wonderfully named Kraken Mare is 1,000 kilometres across, but a journey across this sea by any future human explorers would be fraught with danger. Compared with the liquid water oceans on Earth, the hydrocarbon seas of Titan are much less dense (only half that of water). Ships manufactured on Earth would sink helplessly if they tried to navigate this particular sea.

Titan is a world of two oceanic levels. Deep below its surface, there is a water ocean that was detected by *Cassini*.[148] As Titan orbits Saturn it is squeezed by the gravity of that giant planet. The way Titan distorts is revealed by changes to its gravitational field, and this reveals its internal structure. These changes were detected via changes in the speed of *Cassini* as it made flybys of the moon. The measurements show that Titan is susceptible to tidal effects, indicating a layer of liquid about 100 kilometres below its surface. This water below the surface may be the source of the methane in Titan's atmosphere. Indeed, something must be replenishing the atmospheric methane, as evidence from isotopes of nitrogen around it suggest that Titan has lost something like five times the present amount of its atmosphere to space over its life. Where is the methane coming from? Titan's atmosphere contains a telltale signal in the gas argon, more specifically the isotope ^{14}Ar, which originates from the decay of radioactive potassium in the rocky core of Titan below its ice-water mantle. The methane from such a deep source may then find its way to the surface via cryovolcanoes, in which the 'magma' is water ice warm enough to slowly flow (much as glacier ice flows on Earth) to this moon's surface.

Titan is clearly active geologically. The dramatic scenery carved by the rare, torrential downpours that punctuate the steady drizzle are well preserved, and scarred by very few meteorite impacts. The surface must constantly be being renewed with some version of ice tectonics as well as ice volcanism. The ice sediment, too, must pile up in strata that, like on Earth, must somehow be buried in sedimentary basins. Here, in pore spaces between the sand-sized ice grains, there will be the solar system's richest hydrocarbon reserves. The northern hydrocarbon lakes lie in the low ground of a moon that is slightly flattened at its poles. These lakes may not be so much sitting on solid ice, but marking areas where a hydrocarbon table (akin to Earth's water table) comes to the surface. Titan will be a tempting target for far-future oil prospectors.

It is a target for prospectors of alien life of the present, too, both in the surface ethane seas and in the subsurface water ocean. In the former, there are plenty of complex organic compounds, including tholins (complex polymeric nitrogen-containing hydrocarbons) produced by photochemically driven reactions in the atmosphere. There is none of that supremely universal solvent, liquid water, at the surface however—it is simply far too cold—and it is a moot point whether Titan's liquid methane and ethane would allow the kind of complex and self-organizing chemistry that has made life possible on Earth. However, the subterranean sea is water-rich, and this moon's active geology has almost certainly enabled some kind of cycling to take place between the interior and the organic-rich exterior. Such a link is a potential key to life.

Also circling around Saturn is the tiny moon Enceladus, discovered by William and Caroline Herschel with their '40 foot' telescope in Slough (see Chapter 1). This moon is just under 500 kilometres in diameter—a similar size to the small American state of Maine, or the European country of Belgium. Despite its diminutive size, it is among the most interesting of all Saturn's myriad small worlds. Speeding past this moon in 2005, *Cassini* discovered something remarkable: water-rich plumes, more than a dozen altogether, shooting material into space from its southern polar terrain and feeding a giant plume that extends thousands of kilometres into space. This water contributes material to one of Saturn's famous rings, the 'E-ring', lacing it with an icy spray containing sodium salts.

There seems to be a pressurized, salty ocean below the surface of this moon,[149] heated by the tidal pressures exerted by the neighbouring moon Dione, and by radioactive elements within the silicate core of Enceladus itself. More importantly, *Cassini* recorded other materials in the watery plumes, including some of the basic chemicals that are needed for life: simple organic compounds, nitrogen, and methane (which, just perhaps, might be a product of biological activity). The watery world that is likely present below the southern polar region of Enceladus is seen as one of the best prospects for extraterrestrial life.

Beyond Saturn lie the last of the official planets, Uranus and Neptune. Both are regarded as 'ice giants', named after mythological characters (Jupiter's grandfather and the Roman god of the sea respectively, although Uranus only just escaped being called Georgium Sidus, after the famously mad King George III). In both planets, small metallic and rocky cores are overlain by thick icy mantles of water, ammonia, and methane, in turn covered by dense atmospheres of hydrogen, helium, and ammonia—the last of these giving them their beautiful blue colour. There is plenty of water here, but nothing that we would regard as oceans. Both, though, have many moons, mostly tiny rock and ice aggregates, but some with indications of geological activity driven by past tidal heating. Circling Neptune is the largest of these, Triton, which has a thick icy crust with cryovolcanoes and evidence of a complex geological history, on which geysers of liquid nitrogen have been glimpsed. Possibly, beneath that thick crust, a water ocean lurks there too.

Oceans Unborn

Far out on the edges of our solar system lie the trans-Neptunian objects, the first of these being the icy bodies in the Kuiper Belt. Out here signals from the Earth's radio transmitters, travelling at the speed of light, take some eight hours to arrive. The Kuiper Belt is a giant doughnut-shaped structure made up of many small planetesimals, the largest being the ex-planet Pluto (which has three moons of its own), and these are probably made of methane, ammonia, and water ice. Also in the Kuiper Belt lies a rock/ice body (considered by some a dwarf planet) named 50000 Quaoar, after the creator god of the native Tongva people of western North America (a god who is also called Chingichngish, so in this instance the astronomers were merciful). Quaoar is a little over 600 kilometres across, and recently surprised scientists by giving the spectral signal of crystalline (rather than amorphous) ice at its surface, which suggests an apparent interior temperature of at least −170 degrees Celsius. This is terribly cold, but warmer than had been predicted and, if not enough for a deep-lying ocean, may well allow cryovolcanism of soft, ammonia-laced ice.[150]

Beyond the Kuiper Belt is the 'Scattered Disc', sparsely populated by yet more icy planetesimals. Farther out still, and now just reached by Earth's most distantly travelled spacecraft, *Voyager 1*, the edge of the solar system may lie in the hypothetical Oort Cloud. Nearly one light year from the Sun, this cloud of (hypothesized) comets is thought to extend to a quarter of the way towards our nearest star, Proxima

Centauri. Occasionally perturbed by a passing star, the Oort Cloud may be the source of the comets that race towards the inner regions of the solar system.

So, in our solar system there really is water everywhere, and a diversity of oceans, some larger than those of Earth. Let us count them up. Including Earth, there are currently seven (eight if we count Titan's two-storey ocean system, with hydrocarbons above and water below). Most are subsurface, separated from the exterior by an icy crust: only Earth and Titan possess surface seas, of water and hydrocarbons respectively. They vary in temperature and chemistry and physical structure, and most remain profoundly mysterious in many aspects, being sensed only indirectly by the modern magic of spacecraft-borne instrumentation. At least two planets, Venus and Mars, have lost their oceans—not counting any short-lived hydrospheres attached to the planetesimals that collided and aggregated to form the existing planets. More might acquire or reacquire them when the Sun eventually inflates into a red giant (Mars briefly, as we have seen, and perhaps with more relevance for future biology, distant Titan). At least three of the non-Earth oceans (Europa, Titan, and Callisto) and one of the dead ones (Mars) have been linked with the possibility of alien life. That is just the tally for one normal planetary system circling around a small-to-medium-sized, perfectly banal star.

Planets and Kuiper Belts have already been detected around other nearby stars. There must, then, be oceans on worlds so distant from us that even their existence has only just been detected by the most sophisticated of modern techniques. But even this far out in space, we can detect the faintest signals of water, and can use that to begin to reconstruct oceans in distant star systems.

10

Undreamed Shores

> A cause more promising
> Than a wild dedication of yourselves
> To unpath'd waters, undream'd shores
>
> William Shakespeare, *A Winter's Tale*

Giordano Bruno came to Oxford University in 1583 hoping to gain a position there—alas unsuccessfully. Had he succeeded, then he would certainly have livened debate at a seat of learning that was 'celebrious', even then. Perhaps he would have escaped, 17 years later, a terrible death at the hands of the Italian Inquisition. He was burned at the stake at the incongruously named Campo de Fiore ('field of flowers') in Rome for holding, and not recanting, a whole range of heresies. Among these was a vision of an infinite universe with many stars around which circled many planets. In Bruno's infinity of worlds, the Earth was neither alone nor special.

Bruno's vision is developed in *On the Infinite Universe and Worlds*, written in 1584, which is certainly not a work of science, but rather one of polemic and philosophy set out as a series of dialogues between Philotheo—a lightly disguised Bruno—and his admiring questioners Elpino, Fracastoro, and Burchio. It is (be warned) heavy going. But his intuition of a universe 'with neither Centre nor Circumference' is clear, and strikingly modern:

> There is a single general space, a single vast immensity which we may freely call void: in it are innumerable globes like this on which we live and grow, this space we declare to be infinite, since neither reason, convenience, our perceptions nor nature assign a limit to it. In it is an infinity of worlds of the same kind as our own.

Bruno's imagination and free thinking led him down many paths, such as contemplation of pantheism and questioning of the Christian Holy Trinity (rather worse crimes than his astronomical heresies, in the view of the Inquisition). He considered the nature of memory, and developed a sophisticated mnemonic system for enhancing it (his subsequent feats of memory winning him considerable celebrity). His intuitions were not always on target (for instance he considered that the atoms that made up matter were alive, and possessed intelligence). Nevertheless, he was probably a genius, or close to it.

He was, too, his own worst enemy. Tactless, stubborn, and abrasive, throughout his life Bruno managed to alienate a good many people who were aware of his talents and who might otherwise have helped him. At Oxford, the dons soon put a stop to his lectures, partly because of their subject matter and partly because of his combative style.[151] The then Master of college later drily wrote of 'that Italian Didapper'[152] who 'more boldly than wisely got up into the higher place of our best and most renowned school, stripping up his sleeves like some juggler, and...undertook among very many other matters to set on foot the opinion of Copernicus that the earth did go round and the heavens did stand still'. So Bruno moved on, and on, until his fateful appointment with the Inquisition.

Since Bruno's time, understanding of the universe has grown apace. His central vision of the immensity of space, with its countless stars, has come to be the one we hold today. However, while stars and distant galaxies have long been observed, the presence—or absence—of distant worlds around those stars has remained a matter of speculation, almost up to the present day. Our own solar system, with its inner rocky planets—only one of which is adorned with surface oceans of liquid water—and its outer gas and ice giants, with their stable, near-circular orbits, was long the only available model of how a planetary system forms. Is our solar system the norm—or a rarity? No one knew.

The problem is that stars are so bright and large and far away, and planets within their glare are so small and dim. Until recently, even the best telescope could not distinguish them. It has taken almost superhuman levels of skill, ingenuity, and patience to separate the tiny signals of planets from those of their parent stars. Almost yesterday, it seems, this was first achieved. Now it is as if a dam has burst, and astronomers are deluged with ever-increasing reports of new exoplanets (extra-solar planets). The universe has become, almost overnight, a much more crowded and various place: one which is downright *weirder* than even science fiction writers had imagined. Furthermore, some of these planets are clearly water-bearing worlds.

The Planets Beyond

The first detection of a planet beyond our solar system came not from around a normal or 'main sequence' star, but from around a pulsar. Pulsars are rapidly rotating neutron stars (the orbital period can be much less than a second), which are the shrunken and hyper-dense remnants of a star after it has become a supernova. Pulsars emit narrow beams of electromagnetic radiation that, on sweeping round to shine on Earth, produce pulses of radiation so regular that they were initially thought to be signals from an alien intelligence. Hence the first pulsar was, only partly jokingly, designated as LGM-1, with LGM standing for 'little green men'. The signal is so regular that any tiny deviations in it can be used to infer the presence of planetary bodies orbiting around the pulsar.

In this way, in 1992, the Polish astronomer Alexander Wolszczan and his Canadian colleague Dale Frail, working at the Arecibo Observatory in Puerto Rica, discovered a planetary system around the pulsar PSR B1257+12, which lies nearly 1,000 light years from Earth in the direction of the constellation Virgo.[153] They detected two planets orbiting their pulsar with periods of 66.6 and 98.2 days respectively (they speculated on a third, later confirmed, and there may be a fourth). These 'pulsar planets' came as a surprise, for it was thought that a supernova should blast away everything in its path.

Then, in 1995, came unequivocal evidence of a planet orbiting a yellow main sequence star—a star more like our own. Fifty light years away in the constellation of Pegasus,[154] 51 Pegasi b was discovered around its star. The scientists who found it, Michel Mayor and Didier Queloz of the Observatoire de Haute-Provence, did not see it, any more than Wolszczan and Frail could have seen the planets reflected in the remaining faint light emitted by pulsar PSR B1257+12. Rather, they observed that the parent star was wobbling very slightly, as it was tugged by the gravitational effect of its orbiting planet. The gravitational effect of the planet produces changes in the position of the star relative to the Earth. Even across the vastness of space those changes can be detected, using a spectroscope, by the Doppler effect. This is the same effect that causes the siren of an oncoming ambulance to sound higher-pitched as it approaches, as the sound waves (to a stationary observer) become more closely spaced, and deeper-toned as the ambulance recedes, the sound waves then 'stretching out'. When the wobble of the star moves it towards the Earth its visible light spectrum is blue-shifted by the Doppler effect, while it is red-shifted as the star moves away. It was the wobble technique (radial velocity, it is called more formally) that Michel Mayor and Didier Queloz used to detect Pegasi b. This technique not only shows the presence of a planet, but also gives information on its mass, on how frequently it orbits its sun (its periodicity), and on how elliptical that orbit is.

Another surprise: Pegasi b turned out to be a giant planet, comparable to Jupiter. But it orbited much more closely to its sun, just 7 million kilometres away, in a tiny orbit that gives it a very short year of just four Earth days. So close to the sun, its surface temperature was calculated to exceed 1,000 degrees Celsius. It turned our own solar system on its head, and so it came to be known as a 'hot Jupiter'—the first of many such, as it turned out.

Exploiting the tiny wobble of planet-bearing stars remains a key way of detecting exoplanets. But other techniques have been invented. One makes use of the transit effect. As a planet passes across ('transits') the face of its star it dims the light of that star by a tiny, though detectable, fraction. The amount of light blocked out by the planet's transit gives a measure of the size of the planet which, if coupled with information about its mass, can provide an indication of its density, and thus what the planet is likely made from. The transit method can also reveal the composition of exoplanetary atmospheres, from spectroscopic analysis of the starlight that passes through them.

Planets can also influence the bending of light itself. Einstein's theory of relativity successfully predicted that a path of light will be bent by the gravity of a massive object such as a star or galaxy. In 1919 the British scientist Arthur Eddington (1882–1944) travelled to the island of Principe off the west coast of Africa to observe the solar eclipse of 29 May that year. Eddington, a Quaker pacifist as well as a brilliant physicist, had only just escaped imprisonment during the First World War for his firmly held pacifist and internationalist views: he was a conscientious objector, and advocated continuing dialogue with German scientists. After the War, he brought Einstein's theory of general relativity to an English-speaking audience (he was an effective popularizer of science as well as being, reputedly, one of 'only three men in the world' who then understood Einstein's revolutionary and counter-intuitive ideas).

At Principe, Eddington was to provide crucial support for Einstein's seemingly topsy-turvy theory. He noted that stars whose light was passing close to our Sun appeared slightly out of position, indicating that the mass of the Sun had bent the light from those stars on its journey to Earth in accordance with Einstein's theory (we now know that light can even be swallowed entirely, in the massive gravity well of a black hole). This same microlensing effect can be used to detect exoplanets, but it needs the light of two stars: that of the background star is bent by the gravity of the intervening star, and if the intervening star has an exoplanet with its own gravity, then that produces a tiny but distinct and detectable change to the microlensing effect.

The Space Race

As the wobble, transit, and microlensing effects were used by ground-based telescopes, with ever more careful and sensitive techniques the number of suspected and confirmed exoplanets began to rise. But the Earth's atmosphere, with its dust and clouds and the absorptive effect of the gases within it, blocks or distorts much of the electromagnetic signal that comes from outer space. In particular, this atmospheric blurring made it virtually impossible to detect Earth-size planets that have Earth-size orbits. One answer has been to go into the transparent near-vacuum of space.

The celebrated pioneer here is NASA's Hubble Space Telescope, launched in 1990. Hubble made a famously inauspicious beginning. A tiny flaw in its 2.4-metre light-collecting mirror meant that its initial images were slightly blurred: they were better than images that could be resolved on Earth, but not as good as the perfect images that were expected. In a magnificent exercise in rescue engineering, NASA sent a mission into space during December 1993 to fix the problem, fitting a series of small mirrors that corrected the problem. With its range of light-collecting instruments that see in ultraviolet, near infrared, and visible light, Hubble has gazed into the depths of space, revealing the age of the universe, the birth of stars—and it has detected a few distant planets too. But Hubble is an all-purpose telescope. Finding planets is specialist work, and two purpose-built space telescopes have now been sent into orbit in search of alien planets. They have found them—in their hundreds.

The European Space Agency's COROT satellite was launched at the end of 2006, and began collecting data in early 2007. Its name stands for 'COnvection ROtation et Transits planétaires' and seems only by chance to be synonymous with the great pre-Impressionist painter Jean-Baptiste-Camille Corot (1796–1875). Corot seems to have no connection with astronomy,[155] but perhaps the name is apt. In his delicate, unshowy landscapes and portraits, he followed the advice of his teacher, Achille-Etna Michallon, to 'render with the utmost scrupulousness everything that I saw before me'. Both COROT and Corot have shown, indeed, an exquisite sensitivity in the delicate capture of light.

In March 2009, the Kepler space telescope was launched by NASA. Here the attribution is quite clear and non-acronymic. This mission was named after Johannes Kepler (1571–1630), the great German mathematician, astronomer, and (as was common in those days) astrologer. Kepler was an early defender (along with Bruno and Galileo) of the Copernican model of the solar system, introduced physics into astronomy, and discovered that the orbits of the planets of our solar system were eccentric, not circular.

Both COROT and Kepler were designed specifically to look for exoplanets by means of the transit method, scanning areas of space in which there are hundreds of thousands of stars and looking for the telltale slight, regular dips in brightness as a planet comes between the parent star and the orbiting telescope. The job is not finished at that point. There is a good deal of cross-checking involving both other satellites and ground-based telescopes, because the dips in luminosity may be due not to planets, but to (say) the less luminous member of a pair of binary stars. It is best to combine the transit data with those of gravitational wobble and, if possible, microlensing too (although even in the star-filled heavens it is rare that one star lies *exactly* behind another).

The inventory of confirmed planets[156] out in space grows rapidly. By the summer of 2013, from Kepler alone there were 134 confirmed planets in 76 star systems—but also in excess of 3,000 exoplanet candidates, still to be verified. It is a tricky task, as the small dips in starlight can be confused with inherent fluctuations in the brightness of stars.

Kepler's mission came to an end in 2013, after two of the spinning wheels that kept it in position failed (the craft might still be put to use to gather other data, however, even in its crippled state). COROT, too, no longer functions. Like Kepler, it lasted a little longer than planned (although for part of that time with only half its information-gathering capacity), then suffered a final computer malfunction late in 2012. A smaller machine than Kepler, it nevertheless discovered over 30 exoplanets, with data from a couple of hundred more candidates still being pored over.

What a variety these exoplanets show, although we hasten to add that the variety we see today is a heavily biased one, even as the data continue to flood in. It is easier to detect big planets than small ones, and those that are close to their parent star than those that are far away. Nevertheless, a pattern is emerging that suggests that our solar system is by no means the standard model—and therefore that the nature and chances of oceans and life (as we know them, at least) will be different, out there.

Nicely near-circular orbits are not part of a universal planetary groundplan. Many of these planets have freakishly looping orbits. 'hot Jupiters' are not uncommon—although perhaps they are not as relatively abundant as they seemed in the early days of exoplanet-finding. Being the biggest type of exoplanet, and the closest to their sun, they are, after all, also the easiest to detect.

There is another category that is looming large. These are the 'super-Earths', planets that are bigger than Earth but smaller than Neptune. At the moment, they seem to make up somewhere between a third and half of all planets found—and they are mostly in close and tight orbits around their parent star. This has perplexed the planetary theorists a little (a state that has become depressingly familiar to theorists since the first exoplanet was found). It had been thought that there was just not enough rocky material close to a star to grow anything much bigger than an Earth, while farther out, beyond the snow line, planets should quickly grow into gas giants.

To explain the hot Jupiters, theory was amended to allow them to form in distant regions and migrate in towards its star. As for the super-Earth-sized planets, the close-orbit zone, it was thought, should have been swept clear of such bodies. In fact, it seems to be crowded with them. When the mass of those super-Earths was measured (by means of the wobble technique) and combined with the size as measured by the transit technique, it was clear that many of those closely orbiting super-Earths have low densities—that is, they have small rocky cores surrounded by gas. Did they form in the outer regions as 'failed Jupiters' before migrating inwards—or were they assembled in place, close to the star? It is one of many open questions[157] at the beginning of this, our golden age of planetary science.

The diversity of this menagerie is astounding, and, indeed, akin to science fiction. A planet has been found that orbits a binary star system, like Tatooine in *Star Wars*.[158] A planet larger than Jupiter has been detected closely orbiting a star estimated at between just 8 and 10 million years old, a star so young that there is still a circumplanetary disc of gas and dust not yet blown away by the stellar wind.[159] There is, too, the planetary system 'that should not exist':[160] HR 8799, with four giant planets, each much bigger than Jupiter with orbits stretching out to ten times that of Jupiter from our Sun. This is also difficult to fit into planetary theory. Just as the moons of our own solar system turned out to be much stranger and more diverse than originally thought, so too are planetary systems far from ours. And there will be more surprises to come, for sure. Greg Laughlin, an astronomer at the University of California in Santa Cruz, thinks that the planetary theorists will be revising their models for some time to come. Whatever the next big thing in planetary science will be, he said, we won't see it coming.[161]

On the basis of what is known—and it is very early days yet—how many planets might there be in our galaxy, the Milky Way? (It is still quite impossible for us today to peer at planets in other galaxies.) The wobble and transit techniques, so eloquent in many ways, are not very good at answering this question because they are biased towards the discovery of planets that are relatively close to their stars. Microlensing, on the other hand, although based on a rare phenomenon (two stars and the Earth being exactly in line) is not affected by such a bias.

Microlensing can detect planets that are very far from their parent stars and also planets that are 'unbound'—that is, planets that have drifted off to float freely in outer space, as cold, dark, frozen, and certainly lifeless worlds. There turn out to be many of these distant or 'unbound planets' (one thinks of these objects—now forever dark and cut off from the warmth of a sun—with a shudder, a little as one might think of the undead in a horror film). They are as numerous, at least, as are the stars in our galaxy.[162] Microlensing has also been used to suggest that, on average, each star in the Milky Way has at least one 'bound' planet. Planets, it seems, are a rule and not an exception in our galaxy[163]—and presumably in other galaxies too.

Searching for Mirror Earths

If exoplanets are mainly dominated by super-Earths, hot Jupiters, and giant Jupiters—what of ones that might resemble our Earth? Ones, that is, that might not be too large to grow into gas giants, or too small for oceans to be retained. Planets that might not be too hot, so that any water just boils away, or too cold, so that the water just freezes and becomes, in effect, a rock. We are searching here for those 'Goldilocks planets' that might, possibly, harbour water- and carbon-based life as we know it.

Such planets seem rare. A plot made in 2011 of the 650 planets then known[164] showed a cluster of hot Jupiters all closer to their stars than we are to the Sun, a cluster of cold Jupiters orbiting farther out, and a cluster of super-Earths all of which were larger than the Earth and orbiting closer to their star. The Earth, shown on the same plot for comparison, sat in splendid isolation in one corner of the graph.

One reason for this may be the difficulty in detecting such small, distant objects. Alternatively, Earth-like planets might be relatively rare in our galaxy. It is too early to tell. However, since that plot was made Earth-sized planets have been discovered, unearthed by the penetrating eye of the prolific Kepler space telescope. So—even if rare—Earth-style bodies are at least present out there.

Typically, when one has been waiting so long to find Earth-sized planets, two turn up at once,[165] in the same star system (much like Venus and Earth around our Sun). Moreover, they orbit a Sun-like star, now christened Kepler-20, which lies about 1,000 light years away from our planet. There are at least five planets in the system: two Earths and three super-Earths. They are not, though, at Venus-

and Earth-like distances from their star. All five planets have closer orbits than has Mercury to our Sun—and most (including the Earth-sized ones) orbit much closer. These planets are Earth-sized, but almost certainly not Earth-like. They are dry and roastingly hot and, therefore, quite dead.

The envelope of detection continues to be pushed further. There are now several confirmed exoplanets even smaller than Earth, including one, a mere 33 million light years away, which has just been discovered, quite accidentally, by the Spitzer Infra-Red Space Telescope.[166] This telescope is usually used to gather further information on exoplanets already discovered, and it was obtaining data on a Neptune-sized exoplanet around the red dwarf star GJ 436 when it discovered an extra 'dip' in the electromagnetic signal, which turned out to be a 'first' for the Spitzer: an extra, and very small, planet two-thirds the size of Earth. Again, it has a very close orbit (with a year of 1.4 Earth days): so close that rock is likely molten at its surface. The same state is likely for Corot-7b, which may be anything between two and eight Earth masses (estimates differ), but is also so close to its parent star that it is likely to have a molten surface. Such exoplanets have been termed 'lava-ocean planets': ocean worlds of a sort—but not as we know them.

Exoplanetary Water

Water, as we have seen, is common throughout the universe, especially in its gaseous and icy forms. That is no surprise, as the H_2O molecule is made from two of the most common atoms in the universe, the ubiquitous hydrogen and the less common (by a factor of 1,000) but still virtually omnipresent oxygen. Water has been detected in the comets, planets, and moons of our solar system, and as a component of the gas and dust clouds of interstellar space,[167] and even in intergalactic space, perhaps injected there by the enormous power of black holes. Water is everywhere. Astronomers, therefore, have been looking for water—and oceans—on the exoplanets that lie beyond our solar system.

The electromagnetic radiation that travels at the speed of light through space gives us information on the size, density, and orbital characters of planets very far from our solar system. It also gives us clues to the chemistry of those distant worlds, as we can interrogate the faint radiation patterns by spectroscopy. For instance, as radiation passes through the Earth's atmosphere, atoms and molecules absorb that radiation in different wavelengths of ultraviolet, infrared, and visible light. This produces an absorption spectrum which can be detected by a spectroscope. The patterns of absorption lines indicate the presence and abundance of oxygen, water vapour, or carbon dioxide, all gases that have clear absorption bands in the thermal infrared. To any alien spaceships that may one day view our own solar system, the absorption spectra of the atmospheres of Earth, Venus, and Mars will appear quite different. Venus and Mars show the clear signal of carbon dioxide, but neither show strong signals of oxygen or water vapour. The Earth, with all three gases, would stand out as a fundamentally different planet, one where life might have evolved. Viewed from outer space, even Earthshine—sunlight that has bounced back to the Earth off the Moon—has been found to reveal a clear spectral 'biosignature'.[168] It follows that if we can collect enough light from distant planets, then it might be possible for us to decipher the atmospheres of those planets too. The greatest promise for this work lies with telescopes (or rather spectroscopes) in space.

The hot Jupiter planet HD 189733 b was discovered in 2005, orbiting a star 63 light years from Earth. This planet is more massive than Jupiter and is in a very close orbit around its star. A year takes only two Earth days, and its gaseous atmosphere is heated to 1,000 degrees Celsius. As the planet transits across the face of its star, light passing through its atmosphere and travelling to Earth has shown the presence of water,[169] using NASA's infrared Spitzer telescope, and methane, as detected by the Hubble telescope.[170]

The technique is delicate enough not only to detect what kind of gas is present, but its motion also. Winds have been detected on the exoplanet HD 189733 b, which is tidally locked so that one side always faces its sun (rather like the Moon facing the Earth). When it transits, it therefore always presents its dark side to us on Earth. The temperatures of that dark side atmosphere signal that heat is being transported from the daytime side to the night side efficiently. This indicates that there must be atmospheric circulation—and so implies the presence of winds on that distant world.[171]

The Supercritical Oceans of 55 Cancri e

The planet 55 Cancri e is another strange and hot world, orbiting so close to its sun that a year lasts just 18 hours. First discovered by the wobble effect in 2004, the infrared light from this super-Earth exoplanet was detected for the first time in 2012 using the Spitzer telescope.[172] Lying 40 light years away from Earth in the constellation of Cancer, 55 Cancri e is a little over twice the radius of Earth and a little under eight times as massive. To us it is a hellish place, so close to its sun that it is tidally locked with the same side of the planet always facing its star, like HD 189733 b. The side facing the sun may reach as high as 2,000 degrees Celsius, suggesting that the planet does not have a substantial atmosphere with winds that can efficiently distribute the heat around the planet (or, it may be that the infrared light that has been detected is from an unexpectedly hot part of the atmosphere).

One might imagine there should be no water at all at such temperatures, on such a planet. But an Earthbound imagination may not be the best guide. The mass and radius of 55 Cancri e are consistent with a planet composed of a rocky core and a thin envelope of light gases: or, more intriguingly, a rocky core with an envelope of supercritical water. If the latter is the case, then 55 Cancri e has the weirdest oceans of all: oceans of something that is not quite water, and not quite steam.

About one-fifth of the mass of 55 Cancri e consists of light materials, most probably including water. If so, this is supercritical water at high temperature and pressure, existing simultaneously as a liquid that is able to dissolve minerals, and as a gas that is able to diffuse through solids. There are some analogies for this kind of supercritical water on Earth. These are mainly to be found in steam turbines, or in the kinds of industrial processes that remove the caffeine from coffee beans by steam heating. More naturally, they may occur in volcanoes deep below the sea. At shallow depths, water issuing from submarine volcanoes blasts its way to the surface as steam. But deep under the sea, 3 kilometres down, the water is under enormous pressure and it emerges as streams of supercritical, mineral-rich water—as black smokers, that is—at over 375 degrees Celsius. Entire oceans may exist in this form on 55 Cancri e.

Water Worlds

55 Cancri e is a water world, of a sort. Earth is not a water world and never will be. Even if the world warmed sufficiently to melt all of the ice of Antarctica and Greenland, of the Himalayas and the Alps and the Cordillera, sea levels would rise by only 70 metres. This is enough to inundate many of the world's great cities, granted: London, Shanghai, and New York would all be beneath the waves. But it would by no means submerge the world completely. On Earth, water accounts for just some 0.05 per cent of Earth's mass. There are worlds, though, where more than 10 per cent of the mass may be water, and where oceans may be hundreds of kilometres deep. Their abyssal depths

would be so deep and dense that even at high temperatures the pressure would turn the water into ice.

A candidate for such a water world is Gliese 1214 b, lying some 40 light years from Earth in the direction of the constellation Ophiuchus. The planet was initially detected by the dimming effect of its transit across the parent star, a dimming of 1.5 per cent occurring every 1.58 days—and indicating that this planet also has a short year and rapid orbit about its star. Later its wobble was measured, giving the planet's orbit, radius, and mass:[173] Gliese 1214 b is 2.68 times the radius of Earth, and 6.55 times more massive, giving the planet a much lower density overall (1870 ± 400 kg/m^3) than Earth (5520 kg/m^3). That lower density is intriguing for it suggests a world rich in water, a water world of perhaps 25 per cent silicate rock and iron, and 75 per cent water—although there are alternatives, of which one is a planet with a silicate core and a massive envelope of hydrogen and helium.

If it is a water world, though, then its diversity of forms of water would rival that of the Earth. A steam atmosphere would give way to a hot liquid ocean that, as it became further compressed, would turn into a form of 'hot ice'. There are various forms of such solid, high-density water, such as the crystalline 'ice VII', with normal ice being 'ice 1h'. There is also an 'ice IX', a genuine low-temperature, high-pressure form which is not to be confused with the terrifying 'ice-nine' that Kurt Vonnegut invented for his novel *Cat's Cradle*. Vonnegut's thankfully imaginary ice-nine was evoked as a stable form, crystalline at room temperature that, on contact with normal water, could make that water become solid and crystalline too, to threaten satisfyingly dramatic eco-disaster.

Not all the water worlds need be weird and exotic by Earth's standards. One of the goals of the exoplanet community is to find 'Earth twins'—similar-sized planets at a similar distance from their stars, which might possess both oceans and land. The companion planets

Kepler-62e and -62f might, just might, represent such planets. Their distance from us—1,200 light years—makes it hard to say. But they are roughly Earth-sized, and broadly within the habitable zone where liquid water could exist (although in the case of Kepler-62f, the more distant of the two from its sun, a thick heat-trapping blanket would be necessary). Both planets are big enough to retain such an atmosphere though, and the potential similarities with Earth have already triggered speculation of swimming life, and flying life, and potentially intelligent life.[174]

Is this completely untestable speculation? Maybe not completely. Attempts to model how such worlds might evolve suggests that if these are water worlds, then there is a good chance that they are complete water worlds—with sufficient ocean to completely cover the planet's surface. Our own planet might have become like this, after all, with just a slight change in its original recipe (some say more large comets hitting it when our solar system was young). In a total ocean world, marine life might evolve, and maybe even intelligent life—but perhaps not intelligent, manipulative, resource-extracting, and technological life. Think of the whales and dolphins on Earth: intelligent, with sophisticated communication, but without the means to become farmers, engineers, or rocket scientists.[175] Water-covered worlds may well outnumber those Earth-style ones in which there is a nice balance between land and water. Kepler-22b, for instance, was the first planet detected that sits in the habitable zone around its yellow star (see Plate 9). It has a radius more than twice that of Earth but seems less dense, so it might be mainly ocean with a rocky core. Thus, quite extraordinarily deep waters may lie there, quite unlike Earth's oceans, although that has not stopped speculation concerning the possibility of life.

The sum effect of COROT and (especially) Kepler has been to revolutionize our understanding of planets in the cosmos. There are lots of them out there. Most stars have them. The overwhelming majority of them are very unlike Earth, or many are well outside the range of anything we see in the solar system, and many planets have unstable orbits prone to dramatic reorganization. However, if even one planet in a thousand is broadly Earth-sized and within the habitable zone, then there might be as many as a billion potentially habitable planets with plentiful water out there, just in our galaxy. It gives pause for thought.

Where Next?

We can detect ever-increasing numbers of exoplanets and we can speculate from the mass and radius of these planets—and thus from their density—that there are planets with liquid water. We can find water in the atmosphere of a hot Jupiter. As ever, we need to look closer.

There is a monster that is currently growing in the NASA workshops—and eating up most of even that large organization's money. It is the James Webb Telescope. Three times the size of the Hubble, with powers to match, it will see farther, more clearly, and bring the universe into much sharper focus than before.[176] Fingers are crossed that all will go well with the launch (set for 2018). If all does go well, and it arrives safely at 'Lagrange point 2', a million and a half kilometres from Earth, then it can get to work (see Plate 10).

The James Webb is designed as a multi-purpose telescope, like the Hubble, so it will do far more than just look for planets. But its potential is already making exoplanet scientists impatient for the years of waiting to be over. Its giant eye will—*should*—have the capacity to gaze uninterrupted at Earth-sized planets orbiting small suns within the habitable zone, and to 'sniff' the composition of the atmospheres looking for water vapour, carbon dioxide, and gases. Then we will be much closer to detecting distant Earth-style oceans.

Until then, astronomers have to be as ingenious as possible with what remains of the global astronomy budget after the James Webb has had its share. One way is to use hand-me-downs from rich relatives. There is currently an offer of a couple of retired US spy satellites to NASA. They are not ideal for imaging exoplanets (the mirrors, although very high quality, are held in place by struts which, as luck has it 'couldn't be in worse places', as one NASA scientist said).[177] Still, beggars can't be choosers, and already there are plans to make these useful by fitting them with carefully designed coronographs—sunshades designed to block out the light from the star while carefully preserving that from an orbiting planet.

With current technology we cannot see the oceans themselves, for the planets we are looking at are specks of light that are spatially unresolved—meaning you cannot see individual parts of them such as seas, continents, icecaps and so on. But there are telltale signs that can be used: colour, shade, and glint. Because oceans are dark and have different colours from other surfaces such as land, variations in the colour of a planet over time may betray the presence of a liquid water ocean. Oceans are also smooth compared to other surface types, and they can polarize light. Nevertheless, the light can then be scattered by the actions of water vapour and aerosols in the atmosphere, so this is not unproblematic. And there is glint, the specular reflection of light from the ocean not long before the Sun sets. Where there is a large angle between a planet's oceans and its star, then the planet should reflect more light.[178] There are problems with this method too, because light can bounce off clouds in the atmosphere in a similar manner, or it can bounce off polar icecaps on an otherwise dry world such as Mars.[179]

And so the search goes on. As you read this book, new discoveries will already have been made. The only certain thing is that these will be surprises.

Oceans Undreamed

We are getting ever closer to discovering a multiplicity of far-distant ocean worlds, some perhaps life-bearing.[180] Planets, we know, are commonplace in the distant heavens. The next generation of satellites and telescopes will find and examine numbers of Earth-sized planets in the habitable zone, and will glimpse evidence of atmospheres, oceans, and perhaps signs of life itself, via planetary chemistries pushed out of equilibrium.

By then we should also have learned more about the hidden ice-covered oceans of such nearer bodies as Europa, Triton, and Callisto (where liquid water has been smuggled far beyond the normal habitable zone by the power of tidal energy), as well as understanding the history of the oceans that seemingly long ago covered, however briefly, the surfaces of our near neighbours Mars and Venus.

Each of these planets, present and past (relatively), near and far, will have oceans quite as various and complex as our own. We see them now as simple images, cartoons almost—blank canvases to which we hope to add detail. Those far oceans will be stirred by currents and by differences in temperature and chemistry, each in a specific pattern and combination, and each different—some of them *very* different—to the patterns and moods we see in our oceans. There will be different *flavours* of ocean out there too. We cannot imagine that any of those oceans will be of pure water. Rather, they will be complex chemical cocktails of dissolved salts and minerals—and of organic compounds too—some dilute, others more concentrated than even the dense brines of the Earth's Dead Sea.

Those far oceans will not be constant, but will evolve through time as their parent planets and stars evolve, either gradually or suddenly and catastrophically. They will, over geological timescales, change in volume, shape, temperature, and chemistry, some waxing larger, others drying out or freezing as water moves between planetary interior, crustal surface, atmosphere, and outer space. The new astronomy will capture oceans young, middle-aged, and old, being born and dying.

We are on the verge of not just a new chapter in oceanography—or exo-oceanography, if you like—but of setting up an entirely new library of oceans, for the diversity and complexity of cosmic oceans will be beyond anything that we can dream of. Truth will be stranger than both fiction and scientific hypothesizing alike—and that is even before we think of the kind of life forms that may be evolving in those extraterrestrial waters.

As these new seascapes open up in front of us, we are here on Earth at another transition: the likely transformation—and biological impoverishment—of our own Earthly oceans that surely still represent a cosmic jewel, even on this widest of universal canvases. For it seems very likely that, over the coming decades, the oceans of Earth will undergo a transformation the like of which has not been seen for many millions of years. The changes wrought by warming, acidification, overfishing, and pollution threaten to kill off not just many species, but also whole ecosystems—not least the extraordinary biological riches of the coral reefs.

It is still—just—not too late to stop or slow this marine holocaust, and there have been many useful initiatives, both national and international. The setting up of marine reserves helps sharply depleted fish populations to recover to something like their former numbers. Current discussions on how the International Law of the Sea may evolve have as one central theme the effects—and possible control—of harmful human activities.[181] Handing the control of local marine resources to local communities has been found to be effective in slowing decline.

The real key to our oceans' future, though, is how we as a global human society manage our need for enormous amounts of energy.

Our current reliance on fossil fuels is an addiction in the precise meaning of the word. If we suddenly stopped using coal, oil, and gas there would be mass poverty and mass starvation; many people would die—perhaps most of the human population. That is because we have not developed the alternatives that currently exist and have been proven to work (nuclear fission, renewables) nor yet invented others (stable and controlled nuclear fusion). Hence, to live from year to year we are hooked on carbon-based energy. If that addiction is not controlled, the oceans will permanently change their character (see Chapter 7).

It would be ironic, in that very human way, to discover a wealth of strange and bizarre oceans out in the cosmos just as we are dismantling the beautiful and unique oceans on our own doorstep. Those distant oceans are, for any foreseeable human prospects, entirely unreachable. Many will be intricate and fascinating as regards their physics and chemistry, but will be biologically dead. Of the living ones that now seem likely to be out there, most will be dominated by microbes—the condition of life in the Earth's oceans for more than three-quarters of their history, after all. Few will have the kind of biological riches of Earthly seas. None of them will suit *us* as well as our own oceans do.

So by all means let us lift an ever more penetrating gaze to the heavens, and enlarge our vision of the many forms that oceans might take. But we have, on our own planet, oceans that are special and unique—and on which all of us ultimately depend. Their wealth and beauty will disappear if we do nothing and simply carry on with business as usual. Let us wish—and work—for the right kind of sea change.

Notes

Chapter 1

1. Bradford et al. (2011). See also <http://www.nasa.gov/topics/universe/features/universe20110722.html> (accessed May 2014).
2. Good general accounts here are Encrenaz (2008) and Kotwicki (2009).
3. Although not quite as mysterious as dark energy, which makes up most of the universe and is causing it to expand ever faster.
4. Yes, that is much faster than the speed of light, which is one of the mysteries of cosmic inflation.
5. We are not sure how rare. Probably *very* rare.
6. Loeb (2013). See also <http://www.nature.com/news/life-possible-in-the-early-universe-1.14341> (accessed May 2014).
7. Van Dishoeck et al. (2011). See also Bergin & Van Dishoeck (2012).
8. Lunine (2006).
9. Encrenaz (2008).
10. Salmeron & Ireland (2012) explore the kind of processes taking place, and how it affects the space rubble (meteorites) forming in that environment.
11. Slower-moving—that is, relative to one another. They are all whirling rapidly around the new Sun.
12. Albarède, F. (2009).

Chapter 2

13. See for example, Lunine (2006) and Hazen (2012).
14. Pearson et al. (2014).
15. Canup (2013).

16. But note Pieters et al. (2009) reported water on the surface of the Moon: perhaps from comet impacts, or from protons in the solar wind combining with oxygen in lunar minerals. Robinson & Taylor (2014) also summarize evidence for small amounts of original water inside the Moon itself.
17. Goldblatt et al. (2010).
18. Albarède (2009).
19. The composition of carbonaceous chondrites seems to be consistent with them as a possible ocean source, see Alexander et al. (2012).
20. Wood et al. (2010).
21. Morbidelli et al. (2012), for example, suggested that much of the water came in at late stages of accretion rather than as a post-accretion 'late veneer'.
22. There has been recent support for a volatile-rich late veneer from the proportions of sulphur, selenium, and tellurium on Earth, which resemble those of carbonaceous chondrites (Wang & Becker, 2013). However, that veneer might have been patchily applied (Klein, 2011).
23. It does this by emitting a beta particle—an electron—from its nucleus, so causing one of the neutrons to turn into a proton. The nucleus therefore goes from comprising one proton and two neutrons to being made of two protons and one neutron, in the isotope helium-3, which has one less neutron than 'normal' helium-4.
24. These isotopes are formed by different nuclear fusion pathways in different stars, so that 'stardust'—cosmic dust derived from interstellar space which has occasionally been found as specks within meteorites—can have proportions of oxygen isotopes vastly unlike anything from our solar system.
25. Hartogh et al. (2012).
26. Valley et al. (2014).
27. Valley et al. (2002). See also Sleep et al. (2001).
28. Abramov & Mojzis (2009).
29. Moore & Webb (2013).
30. Olivier et al. (2012).
31. Næraa et al. (2012).
32. Shirey & Richardson (2011).
33. Van Kranendonk (2011).

34. This is still very much at the hypothesis stage, for suggestions regarding plate tectonics very early in the Earth's history are still being made (Turner et al., 2014).

Chapter 3

35. Nakajiyma et al. (2009).
36. 'Short' is relative to the custom of those times. *Cosmos* came out in five volumes.
37. Roger (1962; see also Roger, 1997).
38. A kind of silica- and crystal-rich lava.
39. A quarter of the crew deserted during the voyage because of the rigours endured.
40. Corfield (2003).
41. With two exceptions. Firstly, that any two founding members of AMSOC constitute a quorum and could make any initiative in the name of AMSOC without consulting the other members; and secondly, that any new member of AMSOC automatically became a founding member.
42. See Scripps Institution of Oceanography Archives (undated), Albatross Award of the American Miscellaneous Society, and Knauss et al. (1998). Not to be confused with the Dead Albatross Award of the contemporary alternative music scene.
43. The equally distinguished scientist Roger Revelle had obtained his Albatross Award in 1973 for 'coveting the bird above all other awards'.
44. <http://www.iodp.tamu.edu/publicinfo/glomar_challenger.html> (accessed May 2014).
45. Bluck et al. (1980).
46. Korenaga (2008).
47. Major fault lines in downgoing ocean crust seem to be particularly significant as fluid pathways (Garth & Rietbrock, 2014).

Chapter 4

48. Haber also invented and oversaw the use of poison gas in the First World War; his first wife and a son committed suicide because of this. One can explore Haber's almost unbearably divided life in Charles (2005).

49. Scrivner et al. (2004).
50. Lacan & Jeandel (2004).
51. For example, CIESM (2008), Krijgsman et al. (1999), Ryan (2008).
52. Garcia-Castellanos & Villaseñor (2011).
53. Ryan (2008).
54. Garcia-Castellanos et al. (2009).
55. Knauth (2005).
56. Sanford et al. (2013).
57. The process is more complicated in reality, involving bicarbonate as well as carbonate ions. For a marvellous account of the key process of ocean chemistry, David Archer's *The Long Thaw* (2009) is highly recommended.
58. Zachos et al. (2005).
59. Dickson (2002).
60. We have written about this in *The Goldilocks Planet* (2012).
61. Kerr (2002); Dickson (2002).

Chapter 5

62. The extent to whether Homer's poetry is mostly his, or is mostly the inheritance of many previous poets, remains a hot topic in classical scholarship.
63. Peterson et al. (1996).
64. Rahmstorf (1999).
65. Collectively, we are currently outgunning the Earth's total geothermal output—that is the heat conducted upwards through the Earth's crust—by something like three times. It gives pause for thought.
66. Alley (2007). The title of this paper simply starts 'Wally was right…'
67. Not the Roman navy of Julius Caesar, though. One of the factors behind the limited success of Caesar's attempted invasions of Britain in 55 and 54 BC was the high tides of the British coastline that surprised his navy, used to the tideless Mediterranean seas, and wrecked a number of their ships.
68. Peterson et al. 1996.
69. Anderson & Rice (2006).
70. Hay (2008).

Chapter 6

71. Arndt & Nisbet (2012).
72. And now, mostly, amid towns and cities.
73. Autocatalytic mechanisms occur for example in the Krebs Cycle, and these are essential components of metabolism that may have evolved pre-biotically.
74. Vasas et al. (2010).
75. See Englehart & Hud (2010).
76. Ling (2004).
77. Sleep (2010).
78. DeLong (2003).
79. Tomitani et al. (2006).
80. Knoll et al. (2006).
81. Butterfield (2000).
82. Ratcliff et al. (2012).
83. Yin et al. (2015).
84. de Goeij et al. (2013).
85. Lenton et al. (2014).
86. Howe et al. (2012).
87. Menon et al. (2013).
88. The trilobites, long thought thoroughly submarine, did emerge into the intertidal zone (Mangano et al., 2014).
89. Houben et al. (2013).
90. Nicol et al. (2010).

Chapter 7

91. Pauly (1995); see also Schrope (2006).
92. Myers & Worm (2003).
93. Frank et al. (2011).
94. Amid the adventure, the book paints a detailed picture of the geography and biology of the sea based on the ocean science of its day. No wonder it was a favourite of Jacques Cousteau's.
95. Devine et al. (2006).
96. Myers et al. (2007).

97. Roberts (2007).
98. Smil (2011).
99. Switek (2011).
100. Rochman et al. 2013.
101. <http://www.bbc.co.uk/news/science-environment-25383373> (accessed 2014).
102. Polyak et al. (2010).
103. Fox (2012).
104. Boyce et al. (2010); see also Siegel & Franz (2010).
105. Cheung et al. (2013).
106. Fillippelli (2010).
107. Canfield et al. (2010).
108. Diaz & Rosenberg (2008).
109. Caldeira & Wickett (2003).
110. Bednaršek et al. (2013).
111. For example: the splendid, if dismaying, *Ocean of Life* by Callum Roberts (2012).
112. Alexis Madrigal's *Powering the Dream* (2011) shows (amid providing fascinating context) how much damage *unstable* regulatory frameworks can do.
113. Crutzen (2002); see also Waters et al. (2014).

Chapter 8

114. Haff (2013).
115. Dixon (1981).
116. Bounama et al. (2001).
117. Bounama et al. (2001).
118. O'Malley-James et al. (2012).
119. Kasting (1988).
120. O'Malley-James et al. (2012).
121. If the polar regions continue to exist as relatively colder areas, that is. For the Moon will have been receding farther and farther from the Earth as the incessant action of the tides slowly removed energy from the Earth–Moon system. One billion years in the future, it might have retreated far enough for the Moon's stabilizing effect on the Earth's

spin to have ended, and then the Earth will be free to rotate more chaotically, so that at some times it can almost be rolling like a barrel in its orbit, allowing sunlight to shine much more evenly on the planetary surface.

122. See also Watson et al. (1984), who calculated slightly lower temperatures of around 900 degrees Celsius—still hot enough to melt most rocks and cause others to go plastic. They noted that this will help the oxygen produced become absorbed back into the interior of the planet; as the water molecules are dissociated by rock reactions the hydrogen is then stripped off into space. The concept and term 'runaway greenhouse' was originally coined by the Caltech scientist Andrew Ingersoll in a paper published in 1969.

Chapter 9

123. The work did not originally impress his professors at Uppsala—and he was originally awarded a fourth class degree for the dissertation in which he wrote it up.
124. Arrhenius (1918).
125. Kerr (2012b).
126. Ingersoll (2007).
127. Donahue et al. (1982).
128. Probably quite a bit more, since some deuterium must have been lost as well as most of the hydrogen.
129. Andrews-Hanna et al. (2008); Kiefer (2008).
130. Clifford & Parker (2001).
131. Carr & Head (2003).
132. Di Achille & Hynek (2010); Fairén (2010).
133. Fairén et al. (2010).
134. Woodworth-Lynas & Guigné (2003).
135. Moscardelli et al. (2012).
136. Grotzinger et al. (2014).
137. The evidence here is tiny amounts of trapped gases, the chemistry of which precisely matches that of the atmosphere of Mars, as analysed by the *Viking* space lander.
138. Filiberto & Treiman (2009).

139. McCubbin et al. (2012).
140. Fairén et al. (2011).
141. Baker et al. (1991).
142. Janhunen (2002).
143. Chassefière et al. (2007).
144. Khurana et al. (1998).
145. Schmidt et al. (2011).
146. Named after Christiaan Huygens (1629–1695), who discovered Titan. His accomplishments overall were not quite as extravagant as those of Cassini—but he did invent the pendulum clock.
147. Lunine (2009).
148. Kerr (2012a).
149. Postberg et al. (2011).
150. Stevenson (2004).

Chapter 10

151. McMullen (1986).
152. A water bird that bobs its head up and down. The comparison was not meant to be flattering.
153. Wolszczan & Frail (1992).
154. Mayor & Queloz (1995).
155. He did, though, have a main-belt asteroid named after him as 6672 Corot in 1971.
156. The planets have a variety of names, depending on how they were found and who found them. Unlike the familiar names of our own solar system's planets—Mars, Jupiter, or Uranus for example—exoplanets discovered by ground-based methods are denoted according to the star they orbit. Thus, for the exoplanet '55 Cancri e', 55 Cancri refers to the star—which sits in the constellation of Cancer, and this is suffixed by lower case letters, with 'b' denoting the first planet discovered, and then c, d, e, etc. Those discovered by the space telescopes are named after them, such as Kepler-62e and Corot-1b.
157. Hand (2011).
158. Welsh et al. (2012); Southworth (2012).

159. Setiawan et al. (2008).
160. Close (2010); Marois et al. (2010). These were actually imaged because they are so hot and bright and distant from their sun.
161. Hand (2011).
162. Sumi et al. (2011).
163. Cassan et al. (2012).
164. Hand (2011).
165. Fressin et al. (2012); Queloz (2012).
166. <http://www.nasa.gov/mission_pages/spitzer/news/spitzer20120718.html> (accessed May 2014).
167. Van Dishoeck (2011)
168. Sterzik et al. (2012).
169. Knutson (2007); Sasselov (2008); and see Birkby et al. (2013).
170. Showman (2008).
171. Sasselov (2008).
172. Demory et al. (2012); <http://www.nasa.gov/mission_pages/spitzer/news/spitzer20120508.html> (accessed May 2014).
173. Charbonneau et al. (2009).
174. Becker (2013); <http://www.nasa.gov/mission_pages/kepler/news/kepler-62-kepler-69.html> (accessed May 2014).
175. <http://news.harvard.edu/gazette/story/2013/04/water-worlds-surface/> (accessed May 2014).
176. <http://www.space.com/14708-alien-planets-atmosphere-james-webb-space-telescope.html> (accessed May 2014).
177. Cowen (2013).
178. <http://www.astrobio.net/exclusive/4882/can-astronomers-detect-exoplanet-oceans> (accessed May 2014).
179. <http://arxiv.org/abs/1205.1058> (accessed May 2014).
180. <http://phl.upr.edu/library/notes/habitabilityofthepaleo-earthasamodelforearth-likeexoplanets> (accessed May 2014).
181. Vidas (2011).

References and Further Reading

Chapter 1

Albarède, F. 2009. Volatile accretion history of the terrestrial planets and dynamic implications. *Nature* 461, 1227–33.

Bergin, E.A. & Van Dishoeck, E.A. 2012. Water in star- and planet-forming regions. *Philosophical Transactions of the Royal Society* A370, 2778–802.

Bradford, C.M. et al. 2011. The water vapor spectrum of APM 08279+5255: X-ray heating and infrared pumping over hundreds of parsecs. *Astrophysical Journal Letters*, arXiv: 1106.4301v2.

Encrenaz, T. 2008. Water in the solar system. *Annual Review of Astronomy and Astrophysics* 46, 57–87.

Kotwicki, V. 2009. Water in the universe. *Hydrological Sciences Journal* 36 (1), 49–66.

Loeb, A. 2013. The habitable epoch of the early universe. *Astrobiology*, arXiv: 1312.0613v1.

Lunine, J.I. 2006. Origin of water ice in the solar system. In D.S. Lauretta & H. McSween (eds), *Meteorites and the early solar system*, Vol. 2. University of Arizona Press.

Salmeron, R. & Ireland, R.R. 2012. Formation of chondrules in magnetic winds blowing through the proto-asteroid belt. *Earth and Planetary Science Letters* 327–8, 61–7.

Van Dishoeck, E.F. 2011. Water in space. *Europhysics News* 42 (1), 26–31.

Van Dishoeck, E.F. and 70 others. 2011. Water in star-forming regions with the Herschel Space Observatory (WISH): overview of key program and first results. arXiv: 1012.4570v1.

Chapter 2

Abramov, O. & Mojzis, S.J. 2009. Microbial habitability of the Earth during the late heavy bombardment. *Nature* 459, 419–22.

Albarède, F. 2009. Volatile accretion history of the terrestrial planets and dynamic implications. *Nature* 461, 1227–33.

Alexander, C.M. et al. 2012. The provenances of asteroids, and their contribution to the volatile inventories of the terrestrial planets. *Science* 337, 721–3.

Canup. 2013. Lunar conspiracies. *Nature* 504, 27–9.

Goldblatt, C., Zahnle, K.J., Sleep, N.H., & Nisbet, E.G. 2010. The eons of Chaos and Hades. *Solid Earth* 1, 1–3.

Hartogh, P. et al. 2012. Ocean-like water in the Jupiter-family comet 103P/Hartley 2. *Nature* 478, 218–20.

Hazen, R.M. 2012. *The story of Earth*. Viking Books.

Klein, T. 2011. Earth's patchy late veneer. *Nature* 477, 168–9.

Lunine, J.I. 2006. Physical conditions on the early Earth. *Philosophical Transactions of the Royal Society* B361, 1721–31.

Moore, W.B. & Webb, A.G. 2013. Heat-pipe Earth. *Nature* 501, 501–5.

Morbidelli, A., Lunine, J.I., O'Brien, D.P., Raymond, S.N., & Walsh, K.J. 2012. Building terrestrial planets. *Annual Review of Earth and Planetary Sciences* 40, 251–75.

Næraa, T., Scherstén, A., Rosing, M.T., Kemp, A.I.S., Hoffman, J.E., Kokfelt, T.F., & Whitehouse, M.J. 2012. Hafnium isotope evidence for a transition in the dynamics of continental growth 3.2 Gyr ago. *Nature* 485, 627–30.

Olivier, N., Dromart, G., Coltice, N., Flament, N., Rey, P., & Sauvestre, R. 2012. A deep subaqueous fan depositional model for the Paleoarchaean (3.46 Ga) Marble Bar Charts, Warrawoona Group, Western Australia. *Geological Magazine* 149, 743–9.

Pearson, D.G. et al. 2014. Hydrous mantle transition zone indicated by ringwoodite included within diamond. *Nature* 507, 221–4.

Pieters, C.M. et al. 2009. Character and spatial distribution of OH/H_2O on the surface of the Moon seen by M3 on Chandrayaan-1. *Science* 326, 568–72.

Shirey, S.B. & Richardson, S.H. 2011. Start of the Wilson cycle at 3 Ga shown by diamonds from subcontinental mantle. *Science* 333, 434–6.

Sleep, N.H., Zahnle, K., & Neuhoff, P.S. 2001. Initiation of clement surface conditions on the early Earth. *PNAS* 98 (7), 3666–72.

Turner, S., Rushmer, T., Reagan, M., & Moyen, J-F. 2014. Heading down early on? Start of subduction on Earth. *Geology* 42, 139–42.

Valley, J.W., Peck, W.H., King, E.M., & Wilde, S.A. 2002. A cool early Earth. *Geology* 30, 351–4.

Valley, J.W. et al. 2014. Hadean age for a post-magma-ocean zircon confirmed by atom-probe tomography. *Nature Geoscience* 7, 219–23.

Van Kranendonk, M.J. 2011. Onset of plate tectonics. *Science* 333, 413–14.

Wang, Z. & Becker, H. 2013. Ratios of S, Se and Te in the silicate Earth require a volatile-rich late veneer. *Nature* 499, 328–31.

Wood, B.J., Halliday, A.N., & Rehkämper, M. 2010 (reply to Albarède, F. 2009). Volatile accretion history of the Earth. *Nature* 467, e6–7.

Chapter 3

Bluck, B.J., Halliday, A.N., Aftalion, M., & Macintyre, R.M. 1980. Age and origin of Ballantrae ophiolite and its significance to the Caledonian orogeny and Ordovician time scale. *Geology* 8, 492–5.

Corfield, R. 2003. *The silent landscape: The scientific voyage of the HMS Challenger*. National Academies Press.

Garth, T. & Rietbrock, A. 2014. Order of magnitude increase in subducted H_2O due to hydrated normal faults within the Wadati-Benioff zone. *Geology* 42, 207–10.

Knauss, J., Lill, G., & Maxwell, A. 1998. Recounting the history of the Albatross Award. *EOS, Transactions of the American Geophysical Union* 79(3), 31.

Korenaga, J. 2008. Plate tectonics, flood basalts and the evolution of Earth's oceans. *Terra Nova* 20, 419–39.

Nakajiyma, J., Tsuji, Y., & Hasegawa, A. 2009. Seismic evidence for thermally-controlled dehydration in subducting oceanic crust. *Geophysical Research Letters* 36, L03303.

Roger, J. (ed.) 1962. *Buffon: Les époques de la nature*. Mémoires du Muséum National d'Histoire Naturelle. Nouvelle Série, Série C, Sciences de la Terre, Tome X. Paris, Editions du Muséum.

Roger, J. 1997. *Buffon: A life in natural history*. Cornell University Press.

Chapter 4

Archer, D. 2009. *The long thaw*. Princeton University Press.

Charles, D. 2005. *Between genius and genocide: The tragedy of Fritz Haber, father of chemical warfare*. Jonathan Cape.

CIESM. 2008. The Messinian salinity crisis from mega-deposits to microbiology—a consensus report. No. 33 in CIESM Workshop monographs (edited by F. Brian), Monaco.

Dickson, J.A.D. 2002. Fossil echinoderms as monitor of the Mg/Ca ratio of ancient oceans. *Science* 298, 1222–4.

Garcia-Castellanos, D., Estrada, F., Jimenéz-Munt, I., Gorini, C., Fernàndez, M., Vergés, J., & De Vincente, R. 2009. Catastrophic flood of the Mediterranean after the Messinian salinity crisis. *Nature* 462, 778–82.

Garcia-Castellanos, D. & Villaseñor, A. 2011. Messinian salinity crisis regulated by competing tectonics and erosion at the Gibraltar arc. *Nature* 480, 359–63.

Kerr, R.A. 2002. Inconstant ancient seas and life's path. *Science* 298, 1165–6.

Knauth, L.P. 2005. Temperature and salinity history of the Precambrian ocean: Implications for the course of microbial evolution. *Palaeogeography, Palaeoclimatology, Palaeoecology* 219, 53–69.

Krijgsman, W., Hilgen, F.J., Raffi, I., Sierro, F.J., & Wilson, D.S. 1999. Chronology, causes and progression of the Messinian salinity crisis. *Nature* 400, 652–5.

Lacan, F. & Jeandel, C. 2004. Subpolar mode water formation traced by neodymium isotope composition. *Geophysical Research Letters* 31 (14), L14306. doi: 10.1029/2004GL01947.

Ryan, W.B.F. 2008. Decoding the Mediterranean salinity crisis. *Sedimentology* 56, 95–136.

Sanford, W.E., Doughten, M.W., Coplen, T.B., Hunt, A.G., & Bullen, T.B. 2013. Evidence of high salinity of Early Cretaceous sea water from Chesapeake Bay crater. *Nature* 503, 252–6.

Scrivner, A.E., Vance, D., & Rohling, E.J. 2004. New neodymium isotope data quantify Nile involvement in Mediterranean anoxic episodes. *Geology* 32 (7), 565–8.

Zachos, J.C. et al. 2005. Rapid acidification of the ocean during the Paleocene-Eocene Thermal Maximum. *Science* 308, 1611–15.

Zalasiewicz, J. & Williams, M. 2012. *The Goldilocks planet: The 4 billion year story of Earth's climate*. Oxford University Press.

Chapter 5

Alley, R.B. 2007. Wally was right: Predictive ability of the North Atlantic 'conveyor belt' hypothesis for rapid climate change. *Annual Review of Earth and Planetary Sciences* 35, 241–72.

Anderson, T.R. & Rice, T. 2006. Deserts on the sea floor: Edward Forbes and his azoic hypothesis for a lifeless deep ocean. *Endeavour* 30, 131–7.

Hay, W.W. 2008. Evolving ideas about the Cretaceous climate and ocean circulation. *Cretaceous Research* 29, 725–53.

Peterson, R.G., Stramma, L., & Kortum, G. 1996. Early concepts and charts of ocean circulation. *Progress in Oceanography* 37, 1–115.

Rahmstorf, S. 1999. *Currents of change: Investigating the ocean's role in climate*. Essay for the McDonnell Foundation Memorial Fellowship.

Chapter 6

Arndt, N.T. & Nisbet, E.G. 2012. Processes on the young Earth and the habitats of early life. *Annual Review of Earth and Planetary Sciences* 40, 521–49.

Butterfield, N.J. 2000. *Bangiomorpha pubescens* n. gen., n. sp.: Implications for the evolution of sex, multicellularity and the Mesoproterozoic/Neoproterozoic radiation of eukaryotes. *Paleobiology* 26, 386–404.

de Goeij, J.M., van Oevelen, D., Vermeij, M.J.A., Osinga, R., Middelburg, J.J., de Goeij, A.F.P.M., & Admiraal, W. 2013. Surviving in a marine desert: The sponge loop retains resources within coral reefs. *Science* 342, 108–10.

DeLong, E.F. 2003. Oceans of archaea. *ASM News* 69, 503–10.

Englehart, A.E. & Hud, N.V. 2010. Primitive genetic polymers. *Cold Spring Harbor Perspectives in Biology*, 2 (12), a002196, doi: 10.1101/cshperspect.a002196.

Houben, A.J.P. et al. and the Expedition 318 Scientists. 2013. Reorganization of southern ocean plankton ecosystem at the onset of Antarctic glaciation. *Science* 340, 341–4.

Howe, M., Evans, M., Carney, J.N., & Wilby, P.R. 2012. New perspectives on the globally important Ediacaran fossil discoveries in Charnwood Forest, UK: prequel to. *Proceedings of the Yorkshire Geological Society* 59, 137–44.

Knoll, A.H., Javaux, E.J., Hewitt, D., & Cohen, P. 2006. Eukaryotic organisms in Proterozoic oceans. *Philosophical Transactions of the Royal Society B* 361, 1023–8.

Lenton, T.M., Boyle, R.A., Poulton, S.W., Shileds-Zhou, G.A., & Butterfield, N.J. 2014. Co-evolution of eukaryotes and ocean oxygenation in the Neoproterozoic era. *Nature Geoscience* 7, 257–65.

Ling, G. 2004. What determines the normal water content of a living cell? *Physiological Chemistry and Physics and Medical NMR* 36, 1–19.

Mangano, M.G., Buatois, L.A., Astini, R., & Rindsberg, A.K. 2014. Trilobites in early Cambrian tidal flats and the landward expansion of the Cambrian explosion. *Geology* 42, 143–6.

Menon, L.R., McIlroy, D., & Brasier, M.D. 2013. Evidence for Cnidaria-like behavior in ca. 560 Ma Ediacaran *Aspidella. Geology* 41, 895–8.

Nicol, S., Bowie, A., Jarman, S., Lannuzel, D., Meiners, K.M., & Van Der Merwe, P. 2010. Southern Ocean iron fertilization by baleen whales and Antarctic krill. *Fish and Fisheries* 11, 203–9.

Paine, R.T. 1969. A note on trophic complexity and community stability. *The American Naturalist* 103, 91–3.

Ratcliff, W.C., Denison, R.F., Borrello, M., & Travisano, M. 2012. Experimental evolution of multicellularity. *PNAS* 109, 1595–1600.

Sleep, N.H. 2010. The Hadean-Archaean environment. *Cold Spring Harbor Perspectives in Biology* 2 (6), a002527.

Tomitani, A., Knoll, A.H., Cavanaugh, C.M., & Ohno, T. 2006. The evolutionary diversification of cyanobacteria: Molecular-phylogenetic and paleontological perspectives. *PNAS* 103, 5442–7.

Vasas, V., Szathmary, E., & Santos, M. 2010. Lack of evolvability in self-sustaining autocatalytic networks constraints metabolism-first scenarios for the origin of life. *PNAS* 107, 1470–5.

Yin, Z. et al. 2015. Sponge grade body fossil with cellular resolution dating 60 Myr before the Cambrian. PNAS 112, E1453–1460. doi: 10.1073/pnas.1414577112.

Chapter 7

Bednaršek, N. et al. 2013. Extensive dissolution of live pteropods in the Southern Ocean. *Nature Geoscience* 6, 57–60.

Boyce, D.G., Lewis, M.R., & Worm, B. 2010. Global phytoplankton decline over the past century. *Nature* 466, 591–6.

Caldeira, K. & Wickett, M.E. 2003. Anthropogenic carbon and ocean pH. *Nature* 425, 365.

Canfield, D.E., Glazer, A.N., & Falkowski, P.G. 2010. The evolution and future of Earth's nitrogen cycle. *Science* 330, 192–6.

Cheung, W.W.L. et al. 2013. Shrinking of fishes exacerbates impacts of global ocean changes on marine ecosystems. *Nature Climate Change* 3, 254–8.

Crutzen, P.J. 2002. Geology of mankind. *Nature* 415, 23.

Devine, J.A. et al. 2006. Deep-sea fishes qualify as endangered. *Nature* 439, 29.

Diaz, R.J. & Rosenberg, R. 2008. Spreading dead zones and consequences for marine ecosystems. *Science* 321, 926–9.

Fillippelli, G.M. 2010. The global phosphorus cycle: Past, present and future. *Elements* 4, 89–95.

Fox, D. 2012. Trouble bares its claws. *Nature* 492, 170–2.

Frank, K.T., Petrie, B., Fisher, J.A.D., & Leggett, W.C. 2011. Transient dynamics of an altered marine ecosystem. *Nature* 477, 86–89.

Madrigal, A. (2011) *Powering the dream.* Da Capo Press.

Myers, R.A., Baum, J.K., Shepard, T.D., Powers, S.P., & Peterson, C.H. 2007. Cascading effects of the loss of apex predatory sharks from a coastal ocean. *Science* 315, 1846–50.

Myers, R.A. & Worm, B. 2003. Rapid worldwide depletion of predatory fish communities. *Nature* 423, 280–3.

Pauly, D. 1995. Anecdotes and the shifting baseline syndrome of fisheries. *Trends in Ecology and Evolution* 10 (10), 430.

Polyak, L. et al. 2010. History of sea ice in the Arctic. *Quaternary Science Reviews* 29, 1757–78.

Roberts, C. 2007. *The unnatural history of the sea.* Gaia Press.

Roberts, C. 2012 *Ocean of life.* Penguin Books.

Rochman, C. et al. 2013. Classify plastic waste as hazardous. *Nature* 494, 169–71.

Schrope, M. 2006. The real sea change. *Nature* 443, 622–4.

Siegel, D.A. & Franz, B.A. 2010. Oceanography: Century of phytoplankton change. *Nature* 466, 569–71.

Smil, V. 2011. Harvesting the biosphere: The human impact. *Population and Development Review* 37, 613–36.

Switek, B. 2011. *Written in stone: The hidden secrets of fossils.* Bellevue Literary Press.

Waters, C.N., Zalasiewicz, J.A., Williams, M., Ellis, M., & Snelling A.J. (eds) 2014. *A stratigraphical basis for the Anthropocene. Geological Society of London, Special Publication* 395, 321.

Chapter 8

Bounama, C., Franck, S., & von Bloh, W. 2001. The fate of Earth's ocean. *Hydrology and Earth System Sciences* 5 (4), 569–75.

Dixon, D. 1981. *After man: A zoology of the future.* Granada Publishing.

Haff, P.K. 2013. Technology as a human phenomenon: Implications for human wellbeing. *Geological Society of London, Special Publication* 395, 301–9. doi: 10.1144/SP395.4.

Ingersoll, A P. 1969. The runaway greenhouse: A history of water on Venus. *Journal of the Atmospheric Sciences* 26 (6), 1191–8.

Kasting, J.F. 1988. Runaway and moist greenhouse atmospheres and the evolution of Earth and Venus. *Icarus* 74, 472–94.

O'Malley-James, J.T., Greaves, J.S., Raven, J.A., & Cockell, C.S. 2012. Swansong biospheres: Refuges for life and novel microbial biospheres on terrestrial planets near the end of the habitable lifetimes. *International Journal of Astrobiology* arXiv: 1210.5721v1. [astro-ph.Ep] available at: http://arxiv.org/pdf/1210.5721.pdf

Watson, A.J., Donahue, T.M., & Kuhn, W.R. 1984. Temperatures in a runaway greenhouse on the evolving Venus: Implications for water loss. *Earth and Planetary Science Letters* 68, 1–6.

Chapter 9

Andrews-Hanna, J.C., Zuber, M.T., & Banerdt, W.B. 2008. The Borealis Basin and the origin of the Martian crustal dichotomy. *Nature* 453, 1212–15.

Arrhenius, S. 1918. *The destinies of the stars* (translated by J.E. Fries). G.P. Putnam's Sons, The Knickerbocker Press.

Baker, V.R., Strom, R.G., Gulick, V.C., Kargel, J.S., Komatsu, G., & Kale, V.S. 1991. Ancient oceans, ice sheets and the hydrological cycle of Mars. *Nature* 352, 589–94.

Carr, M.H. & Head, J.W. III. 2003. Oceans on Mars: An assessment of the observational evidence and possible fate. *Journal of Geophysical Research* 108 (E5), 5042.

Chassefière, E., Leblanc, F., & Langlais, B. 2007. The combined effects of escape and magnetic field histories at Mars. *Planetary and Space Science* 55, 343–57.

Clifford, S.M. & Parker, T.J. 2001. The evolution of the Martian hydrosphere: Implications for the fate of a primordial ocean and the current state of the northern plains. *Icarus* 154, 40–79.

Di Achille, G. & Hynek, B.M. 2010. Ancient ocean on Mars supported by global distribution of deltas and valleys. *Nature Geoscience* 3, 459–63.

Donahue, T.M., Hoffman, J.H., Hodges, R.R., & Watson, A.J. 1982. Venus was wet: A measurement of the ratio of deuterium to hydrogen. *Science* 216, 630–3.

Fairén, A.G. 2010. Refilling the oceans of early Mars. *Nature Geoscience* 3, 452–3.

Fairén, A.G., Davila, A.F., Gago-Duport, L., Haqq-Misra, J.D., Gil, C., McKay, C.P., & Kasting, J.F. 2011. Cold glacial oceans would have inhibited phyllosilicate sedimentation on early Mars. *Nature Geoscience* 4, 667–70.

Fairén, A.G., Davila, A.F., Lim, D., & McKay, C. 2010. Icebergs on early Mars. Astrobiology Science Conference 2010, 5467, Lunar and Planetary Institute.

Filiberto, J. & Treiman, A.H. 2009. Martian magmas contained abundant chlorine, but little water. *Geology* 37, 1087–90.

Grotzinger, J.P. et al. 2014. A habitable fluvio-lacustrine environment at Yellowknife Bay, Gale Crater, Mars. *Science* 343, 1242777-1-14.

Ingersoll, A.P. 2007. Express dispatches. *Nature* 450, 617–18.

Janhunen, P. 2002. Are the northern plains of Mars a frozen ocean? *Journal of Geophysical Research* 107 (E11), 5103.

Kerr, R.A. 2012a. Cassini spies an ocean inside Saturn's icy, gassy moon Titan. *Science* 336, 1629.

Kerr, R.A. 2012b. Icy-hot Mercury's water pinned down in the dark. *Science* 336, 150.

Khurana, K.K., Kivelson, M.G., Stevenson, D.J., Schubert, G., Russell, C.T., Walker, R.J., Joy, S., & Polanskey, C. 1998. Induced magnetic fields as evidence for subsurface oceans in Europa and Callisto. *Nature* 395, 777–80.

Kiefer, W.S. 2008. Forming the Martian great divide. *Nature* 453, 1191–2.

Lunine, J.I. 2009. Titan as an analog of Earth's past and future. *The European Physical Journal Conferences* 1, 267–74.

McCubbin, F.M. et al. 2012. Hydrous melting of the Martian mantle produced both depleted and enriched shergottites. *Geology* 40, 683–6.

Moscardelli, L., Dooley, T., Dunlop, D., Jackson, M., & Wood, L. 2012. Deep-water polygonal fault systems as terrestrial analogs for large-scale Martian polygonal terrains. *GSA Today* 22 (8), 4–9.

Postberg, F., Schmidt, J., Hillier, J., Kempf, S., & Srama, R. 2011. A salt-water reservoir as the source of a compositionally stratified plume on Enceladus. *Nature* 474, 620–2.

Schmidt, B.E., Blankenship, D.D., Patterson, G.W., & Schenk, P.M. 2011. Active formation of 'chaos terrain' over shallow subsurface water on Europa. *Nature* 479, 502–5.

Stevenson, D.J. 2004. Volcanoes on Quaoar? *Nature* 432, 681–2.

Woodworth-Lynas, C. & Guigné, J.E. 2003. Ice keel scour marks on Mars: evidence for floating and grounding ice floes in Kasei Valles. *Oceanography* 16 (4), 90–7.

Chapter 10

Becker, A. 2013. Water worlds swell hopes of finding life. *New Scientist* 27 April, 10–11.

Birkby, J.L., de Kok, R.J., Brogi, M., de Mooij, E.J.W., Schwarz, H., Albrecht, S., & Snellen, I.A.G. 2013. Detection of water absorption in the day side atmosphere of HD 189733 b using ground-based high-resolution spectroscopy. *Monthly Notices of the Royal Astronomical Society*. Available at: http://arxiv.org/pdf/1307.1133.pdf

Cassan, A. et al. 2012. One or more bound stars per Milky Way star from microlensing observations. *Nature* 481, 167–9.

Charbonneau, D. et al. 2009. A super-Earth transiting a nearby low-mass star. *Nature* 462, 891–4.

Close, L. 2010. A giant surprise. *Nature* 468, 1048–9.

Cowen, R. 2013. Fresh bid to see exo-Earths. *Nature* 493, 464–5.

Demory, B-O., Gillon, M., Seager, S., Benneke, B., Deming, D., & Jackson, B. 2012. Detection of thermal emission from a super-Earth. *Astrophysical Journal Letters* 751 (2), L28.

Fressin, F. et al. 2012. Two Earth-sized planets orbiting Kepler-20. *Nature* 482, 195–8.

Hand, E. 2011. Super-Earths give theorists a super headache. *Nature* 480, 302.

Knutson, H. 2007. Water on distant worlds. *Nature* 448, 143–4.

Marois, C., Zuckerman, B., Konopacky, Q.M., Macintosh, B., & Barman, T. 2010. Images of a fourth planet orbiting HR 8799. *Nature* 468, 1080–3.

Mayor, M. & Queloz, D. 1995. A Jupiter-mass companion to a solar-type star. *Nature* 378, 355–9.

McMullen, E. 1986. Giordano Bruno at Oxford. *Isis* 77 (1), 85–94.

Queloz, D. 2012. An Earth-sized duo. *Nature* 482, 166–7.

Robinson, K.L. & Taylor, G.J. 2014. Heterogeneous distribution of water in the Moon. *Nature Geoscience* 7, 401–8.

Sasselov, D.D. 2008. Extrasolar planets. *Nature* 451, 29–31.

Setiawan, J., Henning, Th., Launhardt, R., Müller, A., Weise, P., & Kürster, M. 2008. A young massive planet in a star-disk system. *Nature* 451, 38–41.

Showman, A.P. 2008. A whiff of methane. *Nature* 452, 296–7.

Southworth, J. 2012. A new class of planet. *Nature* 481, 448–9.

Sterzik, M., Bagnulo, S., & Palle, E. 2012. Biosignatures as revealed by spectropolarimetry of Earthshine. *Nature* 483, 64–6.

Sumi, T. et al. 2011. Unbound or distant planetary mass population detected by gravitational microlensing. *Nature* 473, 349–52.

Van Dishoeck, E.F. 2011. Water in space. *Europhysics News*, 42 (1), 26–31.

Vidas, D. 2011. The Anthropocene and the International Law of the Sea. *Philosophical Transactions of the Royal Society A* 369, 909–25.

Welsh, W.F. et al. 2012. Transiting circumbinary planets Kepler-34b and Kepler-35b. *Nature* 481, 475–9.

Wolszczan, A. & Frail, D.A. 1992. A planetary system around the millisecond pulsar PSR1257+12. *Nature* 355, 145–7.

Further Reading

Broecker, W. 2010. *The great ocean conveyor*. Princeton University Press.

Corfield, R. 2003. *The silent landscape: The scientific voyage of the HMS Challenger*. National Academies Press.

Kasting, J.F. 2010. *How to find a habitable planet*. Princeton University Press.

Kunzig, R. 2000. *Mapping the deep: The extraordinary story of ocean science*. Sort Of Books.

Roberts, C. 2007. *The unnatural history of the sea*. Gaia Press.

Roberts, C. 2012 *Ocean of life*. Penguin Books.

Summerhayes, C.P. & Thorpe, S.A. (eds) 1996. *Oceanography: An illustrated guide*. Manson Publishing.

Ward, P.D. & Brownlee, D. 2002. *The life and death of planet Earth*. Times Books, Henry Holt & Co., LLC.

Index

ARE DOLPHINS REALLY SMART?

The mammal behind the myth

Justin Gregg

978-0-19-966045-2 | Hardback | £16.99

"Serves as both a rigorous litmus test of animal intelligence and a check on human exceptionalism."

Bob Grant, *The Scientist*

"[T]horough and engaging [Gregg's] writing skills are solid and his observations are often fascinating."

Booklist

The Western world has had an enduring love affair with dolphins since the early 1960s, with fanciful claims of their 'healing powers' and 'super intelligence'. Myths and pseudoscience abound on the subject. Justin Gregg weighs up the claims made about dolphin intelligence and separates scientific fact from fiction. He puts our knowledge about dolphin behaviour and intelligence into perspective, with comparisons to scientific studies of other animals, especially the crow family and great apes.

GREEN EQUILIBRIUM

The vital balance of humans and nature

Christopher Wills

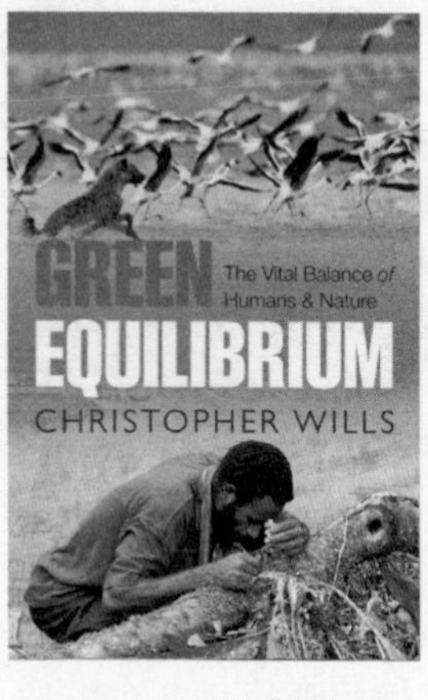

978-0-19-964570-1 | Hardback | £20.00

"*Green Equilibrium* is a richly enlightening exploration of the world and our place in it, infused with fresh insight from science and a deep concern about the future of our planet and our species."

Carl Zimmer, author of *A Planet of Viruses and The Tangled Bank: An Introduction to Evolution*

Across the planet, unique ecosystems are under threat. Overexploitation, pollution, and sheer human population growth put pressure on these delicately balanced 'green equilibria'. Through his vividly described travels, Christopher Wills illustrates the principles of ecology and evolution that underlie the rich mosaics of environments as varied as Californian grasslands, Philippine coral reefs, and the remote mountainous jungle-clad valleys of Papua New Guinea. Using the latest genetic evidence of our evolutionary past, Wills shows how humans form an integral part of the story, being shaped by ecosystems in which we settled as we spread across the planet. Using a number of striking examples he demonstrates how we can halt the damage already done, and help preserve the green equilibria for the local communities who have lived and adapted to them.

THE PLANET IN A PEBBLE

A journey into Earth's deep history

Jan Zalasiewicz

978-0-19-964569-5 | Paperback | £9.99

"A mind-expanding, awe inducing but friendly scientific exploration of the history."

Holly Kyte, *The Sunday Telegraph*

This is a narrative of the Earth's long and dramatic history, as gleaned from a single pebble. It begins as the pebble-particles form amid unimaginable violence in distal realms of the Universe, in the Big Bang, and in supernova explosions, and continues amid the construction of the Solar System. Jan Zalasiewicz shows the almost incredible complexity present in such a small and apparently mundane object. It may be small, and ordinary, this pebble—but it is also an eloquent part of our Earth's extraordinary, never-ending story.

VANISHED OCEAN

How Tethys Reshaped the World

Dorrik Stow

978-0-19-921429-7 | Paperback | £9.99

"*Vanished Ocean* should appeal strongly to legions of former science students who, having since made their way in the world as accountants and personnel managers, hanker for the interest and excitement of a life they once glimpsed but were unable to grasp."

Ted Nield, *Literary Review*

This is a book about an ocean that vanished six million years ago—the ocean of Tethys. Named after a Greek sea nymph, there is a sense of mystery about such a vast, ancient ocean, of which all that remains now are a few little pools, like the Caspian Sea. Dorrik Stow describes the powerful forces that shaped the ocean; the marine life it once held and the rich deposits of oil that life left behind; the impact of its currents on environment and climate.

WAKING THE GIANT

How a changing climate triggers earthquakes, tsunamis, and volcanoes

Bill McGuire

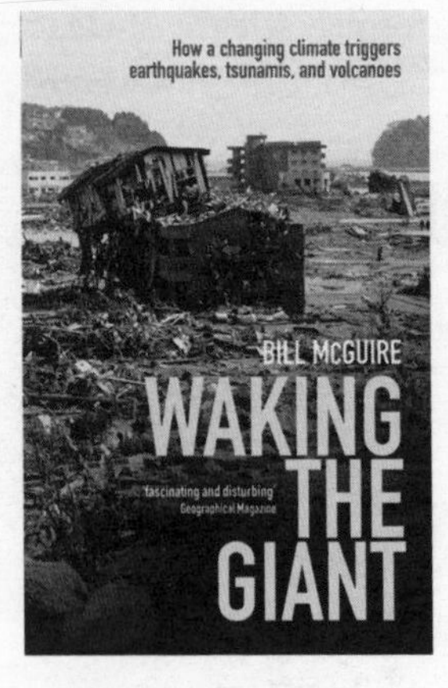

978-0-19-967875-4 | Paperback | £11.99

"McGuire traces this fascinating and disturbing story from the past in order to alert us to present and future perils."

Geographical Magazine

The last twenty thousand years has seen our planet flip from icehouse to greenhouse, provoking earthquakes, tsunamis, and volcanic outbursts. Like a giant stirring from a long sleep, the Earth beneath our feet tossed and turned. Some 15 thousand years ago, ice sheets kilometres thick buried much of Europe and North America, and sea levels were lower. The following 15 millennia, however, saw an astonishing transformation as our planet changed into the temperate world upon which our civilization has grown and thrived. In *Waking the Giant*, Bill McGuire argues that climate change is once more setting the scene for the giant to reawaken. Are we leaving our children not only a far hotter world, but also a more geologically fractious one?